I0016214

On the Road to New Media

Lessons learned in the transformation to digital

Tony Bove
Cheryl Rhodes

Copyright © 2025 by Tony Bove and Cheryl Rhodes. All rights reserved.

This edition was first published in Northern California, the United States of America, in May 2025 by rockument.com.

Printed edition ISBN: 978-0-9989334-3-6

Cover illustration and design: Paul Winternitz

This book was written entirely by humans. We used artificial intelligence (AI) tools solely for research and proofing. For page layout, we used Pages on an Apple Mac.

This book is dedicated to our children.

Acknowledgments

We gratefully acknowledge the help and support of many friends and associates, particularly Denise Caruso and Richard Landry who provided invaluable feedback, and reviewers Marc Canter, Bert Keely, Roger McNamee, Stuart Sharpe, and Paul Winternitz.

Table of Contents

Introduction: A Code You Can Live By

You, who are on the road
Must have a code that you can live by.
— Crosby, Stills, Nash & Young, "Teach Your Children"

L ike many who work in the media, especially in technical writing, we seek the truth. We also want to teach our children the code we used to seek the truth. We wrote this book for the upcoming generations of writers, artists, musicians, performers, designers, publishers, and influencers in the media. You can learn from our successes and failures.

The introduction of the first commercially available microprocessor in 1971 kicked off a torrent of inventions that transformed information from analog to digital, including the internet, video games, artificial intelligence, and social media, and influenced or devoured everything it touched. This book reflects on our points of view over decades of documenting the analog-to-digital revolution. As documentation and technical writing specialists, we were advocates of this revolution, never shying from its grand purpose and theme of doing everything yourself, and doing it in digital form.

The book is a first-person narrative, in which I (Tony) serve as lead narrator, and Cheryl contributes and edits — much like a songwriting team working on lead and harmony vocals, lyrics, and melodies. It chronicles the skills we developed over time, working with new media tools to kickstart pioneering magazines such as *Desktop Publishing*, write over a dozen books about desktop publishing and multimedia including *iPod and iTunes For Dummies*, publish industry-insider newsletters such as the *Bove & Rhodes Report on New Media*, produce the CD-ROM entertainment title *Haight-Ashbury in the Sixties*, and develop apps and documentation for Apple and Android devices.

A very short history

We've kept it short. We would probably drown if we took a broader and deeper plunge into the technological advances of new media. Instead, we offer this personal story of our own long, strange trip down the rabbit hole of digital media, and the key lessons we picked up along the way, so that we might pay them forward to you.

Before the online world existed, the printed book was the preferred medium for communicating deep thought. Civilization now seems to be moving away from text, away even from reading anything of substantial depth. Rather than books, newspapers, and magazines, people now prefer the brief video, the short post, the even shorter tweet, or just a photo, with or without a caption.

Nevertheless, we swim in a sea of media, and our survival as a species depends on whether we can tread the waters of incoherent nonsense — whether we can recognize and reject the false facts and ill-conceived opinions vying for our attention. To recognize truth, the only reliable instruments we have are our thoughts and our history.

To ignore the history of media is to repeat mistakes. The lack of well-developed applications and content doomed many promising media technologies at the starting gate. As Microsoft learned in the 1990s, you can't successfully launch a digital entertainment platform with only the lowest-common-denominator standards in place. The flameouts of web ventures in the early 2000s demonstrated what happens when you go for broke to get "eyeballs" on your screen without a coherent business plan to convert them into customers. And so on.

Turning jobs into opportunities

In the 1970s and 1980s, we wrote printed computer manuals at companies such as Data General and Intel, and worked on computer literacy projects and books about computers. The need for better computer training led us to work closely with software engineers on designing the "help functions" for users of the programs and systems they created. The need for flexible typesetting (to mix fonts for describing step-by-step instructions) led us to experiment with automating typography and composition systems. And the need for peace of mind during the production process drove us to insist on using desktop computers to control the process.

The computer and internet revolutions over the subsequent decades gave us opportunities to start businesses such as magazines and newsletters. We published a magazine as a distribution medium for instructional content, and wrote books for self-training, some of which were best-sellers. However, these revolutions also wreaked havoc with institutional media organizations, as information moved from paper and analog wavelengths into digital formats that could be reproduced flawlessly and infinitely, and archived forever without decay. For those of us who worked in media creation and production, the changes were exciting, but to those who continued to try to turn a profit with paper, the distribution costs were devastating. Magazines sold on newsstands failed and radio all but disappeared, while channels for music and video proliferated without restraint or profit. As classified ad sales plummeted, hedge funds and other financial firms gained control of over half of the daily newspapers in the U.S., and then strip-mined the industry by reducing the number of journalists.

At the same time, opportunities opened up for technical writers and multimedia producers. Teachers of literacy found jobs teaching computer literacy, as we describe in the first chapter. Grammarians became technical editors and project leaders, and news reporters switched their beats to high-tech, as de-

3

scribed in subsequent chapters. Musicians found jobs creating music for games. Video amateurs became professionals at creating how-to videos. Radio shows became podcasts.

In the 1990s we augmented documentation with audio, animation, and video, and distributed the product in the form of multimedia CD-ROM (compact disc — read-only memory) discs. That led rapidly to providing the same multimedia experience on websites, and eventually in mobile apps in the 2000s. A decade later we developed automatically-generated online reference documentation for application programming interfaces (APIs). As we write this in 2025, we have worked in all the new forms of media and publishing of the last five decades.

Ted Nelson, author of *Computer Lib*, summarized our lives nicely when he said "We live in media as fish live in water." His book inspired not only our book but also our careers. Nelson's *Computer Lib* subtitle, "You can and must understand computers now!" was our prime directive. In addition, its style of do-it-yourself (DIY) publishing intrigued us. He used the lowest cost, most affordable print production tools of the day, rather than using expensive professional tools. Despite its unconventional appearance — *Computer Lib* was bound together with a separate volume he authored, entitled *Dream Machines*, in the "head-to-toe" format invented originally to bind the Old and New Testaments of the Bible together), the book effectively and concisely presented the history of computing up to that point, from Nelson's personal viewpoint as a scientist in the field. What was most inspiring was his mix of strong ideas, his higher awareness of how the computer industry works, and his determination to right the wrongs at the intersection of copyright and freedom.

Surviving bad media

This book gives you our perspective on the changes media technology has wrought on our lives and civilization. It also provides tips on how to sharpen your mind to keep ongoing advances in automation, including AI, from eroding your basic mental skills, and how to amplify your own analytical thinking in order to survive bad media. Because one day your car stereo will stop working or the internet will go down, and you'll have to learn how to hear music entirely in your head. Your car's GPS will go out, and you'll wish you still had the paper maps of yesteryear. Your children, raised without the knowledge of history and without training in skills such as driving a car or writing an essay, will be at the mercy of *Terminator* robots.

All this and more will happen, so heed the warnings in this book!

Documenting in Print

Had a love affair but it was only paper. — Talking Heads,
"Paper" (David Byrne)

I n the summer of 1968, at the tender age of thirteen, I watched,
on the CBS television network, scenes from the riot at the
Democratic Convention in Chicago. The police were on a ram-
page, beating up hippies and young liberals alike with billy
clubs, turning crazies like Abbie Hoffman and Jerry Rubin into
heroes. The courageous reporters and photographers of the underground press, a motley collection of alternative newspapers,
political newsletters, and lifestyle broadsheets, were chased and
then beaten by police for covering the event. I learned at an early
age that it might be possible to change the world.

What impressed me most at that time was Paul Krassner's
The Realist, spun out of his office at my favorite publication,
Mad Magazine. Containing heroic articles and reviews alongside
outrageous satire, flagrant sexuality, and rampant conspiracy theories, *The Realist* was my first encounter with a do-it-yourself
underground newsletter operation. I still have fond memories of
the "Disneyland Memorial Orgy" poster and "The Parts That
Were Left Out of the Kennedy Book."

A page from Paul Krassner's The Realist, *courtesy of The Realist Archive Project.*

By the time I reached college I was determined to write and publish a newsletter of my own. As a high-school master of the mimeograph, intoxicated (literally) by its rapturously fragrant, sweetly aromatic pale blue ink, and armed with a typewriter, a pile of stencils, and a tank of nitrous oxide, I pounded out the master pages of a 25-cent alternative literary broadside, titled *Squid Ink*, for the Tufts University class of 1976.

I quickly learned my first lesson in the media business: In publishing you risk your reputation. The stencils did not hide all the spelling and typographic errors that I thought I had covered up with typewriter correction fluid. The final product was an embarrassing mess.

We had not yet met at that time, but Cheryl was more experienced with the mimeograph, having learned how to use it in the first grade of the Wiesbaden AFB dependents school. "I remember how unhappy I was with my mistakes in forming the letter 'W' using a chunky lead pencil with no eraser. When I turned thirteen, I was thrilled to get an inexpensive manual typewriter all my own, the one gift I had requested every birthday and Christmas for several years. But I learned quickly that typewriters did not produce perfectly typed pages that looked like the pages in books."

Typos and misspellings would unwittingly kick us both to move forward into computing and digital media. The first day after graduating college, I came across a *Boston After Dark* issue open to an ad for "creative writing geniuses" that Data General had placed for hiring technical writers. "Write your own ticket," the ad blazed.

That was 1976, the year that Jimmy Carter beat Gerald Ford to become president, the U.S. celebrated its Bicentennial, and nascent personal computer companies such as Apple Computer and Microsoft first incorporated. I needed a job to live in this world.

I had never liked tinkering with radios, electronics, or automobiles, and I had no affinity for computers. I had heard the voice of the computer on the original Star Trek television series, and of course I had seen HAL take over the spaceship in *2001: A Space Odyssey*. But I had no obsession with hardware of any kind. Since it made no sense for them to hire me, I spent the night drinking bourbon and writing a short story about how one of a hundred monkeys pounding on typewriter keys could win a short-story contest, and sent it on with my application. (That idea also surfaced as the hundredth monkey effect in behavioral science, and in *Hundredth Monkey* by Ken Keyes, Jr. published in 1984, but neither was inspired by my short story.)

When the call came from my future employer, I traded my hippie garb for a suit and tie and borrowed a friend's car to get to the interview. The rest is history, and the subject of this book: How we came to understand, embrace, and advocate the use of computers and digital media for creative expression.

In the mid-1970s, the only way to understand and use anything as complex as a computer was to read the printed manuals, with their step-by-step instructions, references, illustrations, and schematics. So as mentioned in the introduction, we set out as technical writers and print-documentation experts to make a living demystifying the mysterious operations behind computers. The projects we encountered, the experiences we shared, and the skills we developed helped us survive the roller-coaster ride into digital media and beyond.

Birth of a symbol

Imagine an entire generation of users having their first encounter with a computer, and not knowing how to proceed. The screen is black, with a small blinking light in the upper corner pulsing in a steady, unbreakable, eternal pattern. The quiet is deafening. It seems to be holding its breath, waiting... for what?

You are supposed to press the Return key, but there is no instruction for it. The Return key is on the far right side of the keyboard, but you don't yet know about it. Even more confusing is the fact that your neighbor's keyboard looks similar, like a typewriter keyboard, but offers an Enter key in the same place as your Return. If this were an imitation of a Federico Fellini film, a monkey would show up out of nowhere and press the Return or Enter key for you.

As technical writers on a documentation team, we argued about using the term *<RET>* for Return. But would the user type angle-bracket **RET** angle-bracket? What would happen on keyboards that show Enter as the key? Do we have to use *<RET> or <ENTER>* at the end of every typed command? Should we instead explain its use in a "Note" in the introduction? What if readers skip the introduction, which they are wont to do? These were the thoughts that tried the souls of early technical writers.

The solution turned out to be a special character: the backward arrow symbol (↵), which was introduced sometime in the late 1970s. You'd see a command like the following:

RUN PROGRAM ↵

Or, if you were playing the original Adventure game – one of the first games designed for a computer – your screen would display:

```
YOU ARE INSIDE A BUILDING, A WELL HOUSE FOR
A LARGE SPRING.

THERE ARE SOME KEYS ON THE GROUND HERE.

THERE IS A SHINY BRASS LAMP NEARBY.

THERE IS FOOD HERE.

THERE IS A BOTTLE OF WATER HERE.
```

Adventure on a CRT terminal. By Autopilot - Own work, CC BY-SA 3.0

What To Do After You Hit Return

One of the first books to be considered "documentation" for computer users was inspired by the "Return" key. *What To Do After You Hit Return*, by Bob Albrecht and the People's Computer Company, covered the nascent computer game industry in 1975.

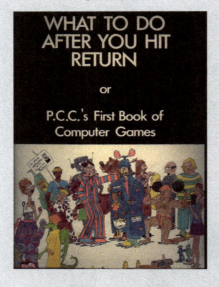

The instructions should show how to type a response, such as:

PICK UP KEYS ↵

Where did this "Return" come from? In pre-computer days, typewriters were equipped with a lever known as a *carriage return*. When engaged, the mechanism pushed or "returned" the paper-holding cylinder ("carriage") to the left side of the paper, and at the same time, rotated the paper upward to begin a new line. When this function migrated to computers, it became two functions: carriage return and line feed (or new line). To initiate both functions, you pressed the Return or Enter key.

> ### *Return or Enter?*
>
> While we prefer "Return" (as it appears on Mac keyboards), the industry previously preferred "Enter" (see Wikipedia) to the point that "Enter Key Day" is celebrated every year on the 18th day of October. For more about this topic, see What are carriage return, linefeed, and form feed? in StackOverflow, a leading source of technical answers, and a trusted online community for developers to learn and share knowledge.

But what technological magic would make it possible to reproduce this backward arrow symbol (⏎) in our printed documentation, when no such key existed on keyboards?

Electric knife and glue

Cheryl and I worked at the minicomputer company Data General (DG) in the late 1970s — I was writing manuals in the software engineering department, and Cheryl was doing the same in the hardware engineering services department.

> ### *Those Data General bastards!*
>
> Data General in the 1970s was the dark horse of the minicomputer industry. (A minicomputer was a computer based on newer processors than the original mainframe computers. Minicomputers were the precursors to personal computers, a.k.a. "microcomputers".) DG was a rude upstart threatening the much larger, more successful, and conservative Digital Equipment Corporation (DEC) down the road.
>
> DG had a very cocky PR department. When IBM entered the minicomputer business, the renegades in PR created an ad, which they circulated inside the company to energize the troops but never ran in a publication, that screamed: "SOME SAY IBM'S ENTRY LEGITIMIZES THE MINICOMPUTER INDUSTRY. THE BASTARDS SAY WELCOME!"

We met during an act of piracy on my part — I was printing multiple copies of a party invitation on a line printer at work, and using an X-Acto knife and Elmer's glue to paste together multiple invites from a single page. She was curious about what I was doing with the shared office printer.

I handed her an invitation to the party, and it was love at first sight. We eventually got together, moved out West together, got married, and had children. But what first impressed me was her attention to detail. She immediately proofread the invitation and pointed out a significant error in the text.

Letters, numbers, and other characters had already evolved into a digital form of text that could be manipulated on and printed from a computer. The American Standard Code for Information Interchange (ASCII), developed from telegraph code, was already a standard all computer manufacturers used to represent text. Computers running typesetting systems were already converting this text into different typefaces and font sizes.

The frank truth about typesetting

Our friend and pioneer Frank Romano, a recognized authority on the history and evolution of type design, technology, and printing, would keep us laughing for hours with his tales of typesetting at the bleeding edge of digital content. For a thorough and illuminating overview of the origins and development of analog-to-digital conversion (which arguably found its first market through computerized typesetting), along with Frank's hilarious anecdotes, see *History of the Phototypesetting Era* by Frank Romano (RIT Press).

At that time Penta (now Penta DTP Publishing) was one of the top suppliers of systems to the typesetting trade. The DG software documentation team used a Penta typesetting system based on DG's Nova minicomputer. The Nova fit into a three-unit 19-inch rack, sported an array of toggle switches for primitive programming, and used disk drives the size of washing machines for data backed up to large reel-to-reel tape machines — like the ones you see in science fiction movies from the 1950s. We learned how to prepare a paper tape to "bootstrap" the machine in order to run its operating system, which controlled the shared disk drives and accessories such as line printers.

Data General was famous for its high-tech disk drives, but at the time they bore more of a resemblance to a Rube Goldberg machine than to the solid-state drives of today. For one thing, the drive lids were prone to flying off due to the vibration of the drive platter's high-velocity vibration speed (think of a top-loading washing machine's spin cycle). The engineering department came up with a solution to this problem that could have rightly been called "putting lipstick on a pig": They got a heavy construction cinder block that was lying around, and placed it on top of the drive to keep it in one piece. Then, somebody actually got the bright idea to include a cinder block with every drive sold to customers. The blocks were painted DG blue, with a DG part number and logo pasted on the side, and were an official part of the drive's bill of materials (which Cheryl managed in the hardware engineering services department).

Hard disk drives consisting of spinning disks coated with magnetic material and a read/write head were relatively new at the time and complemented the magnetic tape drives that archived data, along with the paper tape reader-eaters used to load data. Although tape reels could be written at speed, they were slower to search, and needed to be rewound to retrieve data. The hard disk drive made it possible to control and feed data to phototypesetting machines.

To produce DG's instruction manuals, we turned to skilled phototypesetting operators at the aforementioned Penta, who retyped our words from typewritten pages, using teletype or cathode ray tube (CRT) terminals connected to the Nova. For typewriters, we used the ubiquitous IBM Selectric.

The ubiquitous IBM Selectric

During the 1970s, the IBM Selectric Composer was the most popular typesetting system for small businesses and organizations, featuring magnetic tape memory. Documentation departments typically could not afford one for each writer. The Correcting Selectric II typewriter was far more affordable, but had no memory. Its coolest feature was the ability to immediately correct a typing mistake rather than use cover-up tape, "white-out" correction fluid, or typewriter erasers. Another option was the Wang 1200, which consisted of the logic of a Wang 500 calculator hooked up to a Selectric typewriter for keying and printing, and dual cassette decks for storage. For details, see The Selectric Typewriter and the Wang 1200 word processor.

We would hand stacks of paper produced on our Selectrics over to the phototypesetting operators to be retyped and stored as ASCII text. Scrolls of typeset text rolled out of the typesetting machine and were cut and pasted into pages. We would then have to proofread the typeset text, not just reading for missing sentences or mistyped words, but also hunting for obscure errors such as a word that should have been set in **bold** or `monospace`. This kind of mind-numbing, soul-destroying, tedious proofreading drove a whole generation of young, enthusiastic digital pioneers like us absolutely crazy.

Learning to typeset our own words was our first step in the direction of desktop publishing. It turned out to be a giant leap forward in the evolution of digital media.

Although Wang Labs was just up the road from DG, in an area close to Route 128 that wound around the suburbs of Boston and came to be known as Silicon Necklace, the engineers at DG abhorred the Wang machines for their simplicity, and we just knew that, whether out of sheer Boston pride or competitive disdain, our managers would never have allowed their use. Our documentation team opted to share the Nova and provided computer monitors to the writers. This setup eliminated nearly all of the errors involved with retyping, though not the errors and bad formatting related to typesetting.

The Nova's operating system already included a basic text editor, for use by programmers to write code. The programming task involved constant typing and editing of expressions in a programming language. The text files, often called *source files*, were then compiled into programs. Some features of basic text editors were geared toward programming, such as the find/replace function, which helped programmers find and replace symbol names. Printed output was an afterthought, and in most cases not done at all. There was no need for control over margins, tabs, paragraph indents, or line spacing.

One innovative feature of these early text editors was the ability to move a block of text within a file, or copy the block to another file. These eventually became known as *cut* or *copy* and *paste*, based on the manual steps of cutting a piece of text from the typeset scroll using scissors or a sharp knife, and pasting the cut piece in its proper place on the layout page using contact cement. The idea of cut and paste of print as a composition method was pioneered by the Dada movement and specifically by Marcel Duchamp, the father of conceptual art, in works he called "readymades." *Copy/cut* and *paste* are still in use today as functions in just about every application.

> ### *Xerox popularized cut and paste*
>
> Inspired not only by manual paste-up in production and by early experiments with Douglas Engelbart's mouse device, but also by early line and character editors that separated the move or copy operation into two steps, computer scientist Larry Tesler proposed the names "cut" and "copy" for the first step and "paste" for the second step. Beginning in 1974, he and colleagues at Xerox PARC implemented several text editors that used cut/copy-and-paste commands to move and copy text. Tesler also wrote Pub, a document compiler for producing printable manuals, which was one of the first uses of a markup language.

We also use the terms to describe composing a document by piecing together components of other documents. Serendipitously, we came across a book that had been composed entirely by knife and glue as if it were the product of the Sixties underground press: *Computer Lib* by Ted Nelson would inspire our new advocacy for computers as media tools.

Computer liberation

We introduced *Computer Lib* by Ted Nelson in the introduction, in which he stated unequivocally that "due to ridiculous historical circumstances, computers have been made a mystery to most of the world."

The book's humor was evident from the start. Nelson carried the Dada aesthetic of cut and paste into the merging fields of tech and publishing, and he drew on the Japanese manga binding tradition of slapping two books together upright and reverse of each other — half the book appeared upside-down and began from the back, with the title *Dream Machines*. The cover broke the most basic convention of printed books. The phrase "something is afoot" appeared with an extended cartoon foot on the back-to-front cover, jumping over the formal boundary of the book spine. The book was not just the first desktop-published book — it may also have been the only "floor-top-published" book, as Nelson assembled it on the floor from scraps of typewritten passages, hand-drawn cartoons, and photographs of computer graphics. With missionary zeal he set out to prove that ordinary people could (and must!) understand computers (now!).

Computer Lib *is the legendary bible of the personal computing revolution. The flip side,* Dream Machines, *accurately predicted the impact of interactive computer graphics.*

Nelson's juxtaposition of conventions as far afield as the first printed Bibles and the Dada movement might have struck his first readers as amateurish or even barbaric. In reality, what they were witnessing was the birth of a new aesthetic that broke the rules of commerce to expose their vacuity. (This was the point of the Dada movement too: to bust through the snootiness, serious-ness, and class-consciousness of art criticism in the early 20th century. Duchamp took a urinal and called it a work of art — created by him in the act of labeling it as art, no less — and then just sat back to watch art critics and collectors try to rationalize it in their ruling-class language.)

The 1960s was a time of counterculture rebellion. As Michael Swaine and Paul Freiberger described the early comput-er industry in *Fire in the Valley*, "Only government agencies, universities, and big businesses could afford to own a computer. And they were obscure and sinister, typically operated by a white-coated 'priesthood' of specially-trained operators... com-puters were widely viewed as a dehumanizing tool of the bu-reaucracy... The personal computer was born in a time of social ferment, when idealism ran high... 'Computer power to the peo-ple' was their rallying cry."

Computer Lib had found its mark in Data General's software development department. I rubbed shoulders with a new kind of nerd, far more worldly than the few nerds I'd known in high school. The engineers and technical writers, nearly equally men *and* women, dressed casually, some in the hippie garb of t-shirts, jeans, and sneakers, or even sandals or flip-flops as if their cubi-cles were on the beach. My office mate was a Thomas Pynchon freak who had decorated his side of the cubicle with black-light posters.

It has always been essential for software technical writers to socialize with engineers, who are usually too busy to think about the importance of documentation. There were exceptions in my career, and one was John Gilmore, a future activist for civil liber-

21

ties in cyberspace, who at that time wore a denim sarong. Within a month I found regular customers among the engineers for my Acapulco Gold. I also found that I could detach from the world, dive into the soul of the machine, and write like crazy. I thought I had found the perfect working environment and the perfect job.

At that time companies recruited technical writers mostly from the ranks of engineering graduates who were willing to take the job as a stepping stone to writing code. However, DG took a different approach. My first documentation team was a motley collection of former grammar and English teachers, magazine editors, failed newspaper reporters, and a handful of college graduates like myself.

Writers and engineers socialized together. We were all part of the same engineering department, and treated incursions by marketing personnel as attacks on our frontier. In meetings we would smugly report that progress has been stalled by a near-constant stream of Engineering Change Orders, the aforementioned and dreaded ECOs — more substantial than the disk drive cinder block — that required revisions to the operating manuals.

In short order both engineers and writers learned that it was largely up to the QA (Quality Assurance) department, run by utterly clueless marketing managers, to bless a product offering. The QA teams were mostly recent engineering graduates who couldn't or wouldn't do technical writing.

An engineer, product expert, or marketing manager can't do everything. There always was, and even with artificial intelligence (AI) there will always be, a need for technical writers who can write well, under the mounting pressure of a steadily increasing number of changes to be made before a fast-approaching deadline. And this is another lesson for you: If you want to be gainfully employed, learn how to read critically, write professionally and persuasively, and speak with confidence and clarity before your peers. You may be surprised at how many people can't do these things well.

The art of documentation

For centuries, word of mouth was the only means by which humans could pass on how-to or technical instruction. Early examples of "technical documentation" include explanatory notes written by Copernicus, Hippocrates, Isaac Newton, and Leonardo da Vinci, mostly to demonstrate how to build and use their inventions. Chaucer's *Treatise on the Astrolabe* is recognized as the oldest work written in English on an elaborate scientific instrument.

In retrospect, we are proud that we joined the ranks of the technical writing guild, which had spawned such writers as Thomas Pynchon, Amy Tan, and Kurt Vonnegut.

Technical writing is a very specific type of communication, designed to convey a particular piece of technical information to a particular audience for a particular purpose. Books have been written about technical writing, such as *A Guide to Technical Writing* by T.A. Rickard (1908), and *English for Engineers* by S.A. Harbarger (1923). Joseph D. Chapline wrote a user's manual for the BINAC computer (1949). He became the first technical writer of computer documentation. He went on to document the operation of the UNIVAC computer (1952), using examples to document its functions.

The point of technical writing is to translate technical terms or jargon to simpler language, visualize the "use cases" that customers understand, and describe concisely without sacrificing accuracy or quality, all the while engaging the audience. A good technical writer also must follow the rules for good grammar and style, while also paying attention to brand names and localization rules for translations.

> ### *Learn to write well*
>
> Omit needless words. To be a good writer, or even an excellent writer, follow the instructions in *The Elements of Style* by Strunk and White.
>
> They all boil down to fifty-nine words that, according to E. B. White, could change the world: "Vigorous writing is concise. A sentence should contain no unnecessary words, a paragraph no unnecessary sentences, for the same reason that a drawing should have no unnecessary lines and a machine no unnecessary parts. This requires not that the writer make all sentences short or avoid all detail and treat subjects only in outline, but that every word tell."

A technical writer must also create or follow a documentation style and support the existing means of production. A *style guide*, also known as a manual of style or stylesheet, is a set of standards for the words, formats, and page designs; for example, whether the name of a book is italicized, placed in quotation marks, or underlined; how to format footnotes and references; and whether people must be referred to with gender pronouns (Mr, Ms) or simply by their last names. As writers of magazine articles, books, and manuals, we had to follow the style guide used by the publisher. As publishers of magazines, newsletters, and self-produced books, we have had to devise our own style guides.

Signposts to new spaces

The Chicago Manual of Style is still one of the most widely used and respected style guides in the U.S. for American English publications, including newspapers and magazines. Other countries have their own style guides — in the U.K., for example, see *The Oxford University Style Guide*. The Institute of Electrical and Electronics Engineers (IEEE) style format, widely used for documentation, is based on the Chicago style. The Modern Language Association (MLA) designed the MLA Style for subjects related to the humanities and liberal arts, such as literature, mass communications, and media studies.

A style guide for computer documentation establishes standards to improve user comprehension both within a document, and across multiple documents, such as using a `monospaced` typeface for showing displayed output and programming code, and **bold** for typed commands.

In my role as technical writer in the software department, I would attend development team meetings to get advance notice, and "eat my own dog food" — use the product I was documenting, in order to get realistic examples. It was always good to have another set of eyes probing and testing the product. Most importantly, I would make it known that the documentation was *part of the product*, and act accordingly.

Not only do a company's customers need documentation, they are also attracted to a product through its documentation. In the early days, extensive documentation indicated a company's seriousness, dedication, support, and professionalism. Documentation could help bring in qualified leads for sales, and provide critical information that reduced the cost of support. For example, I knew a company that was spending up to $150 per customer support call until the documentation became available, at which point the company's cost to support customers dropped to $1 per call.

Encoding the seeds of new media

The software documentation team at Data General used computer monitors connected to a DG Nova minicomputer in the nearby lab that ran RDOS (Real-time Disk Operating System) and Business Basic. During this time I wrote the award-winning *Business Basic Reference Manual* on a typewriter. I would compile my examples on the computer and print them out on a line printer. I learned quickly how to use the computer's text editor for writing source code. Eventually we abandoned typewriters and started churning out our documentation in the computer's text editor. RDOS was, among other things, a file management system, so we could transfer electronic files to the typesetting operators who were sharing the Nova.

While solving the problem of reproducing the backward arrow symbol (↵), we learned how to embed typesetting codes in the electronic text to eliminate the typesetting errors and the need for extensive proofreading. We tested this theory with an April Fools edition of the company's internal newsletter, which raised quite a stir among management. This farcical newsletter, full of corny jokes, ironically represented a breakthrough by being the first digitally set publication with our new typesetting system.

DG Mini News April Fools

On April 1, 1977, over a thousand copies of the real Data General Mini News in-house newsletter were replaced with a fake April Fools edition. It looked the same, except that stories proceeded from harmlessly ordinary into the extraordinary, and classified ads veered into the bizarre. Peppered with subtle in-jokes concerning DG's relationship to the community, the edition touted a Monty Pythonesque referral program with cash bonuses up to $5,000, with first referrals for air conditioning engineers at $2,000 and first referrals for technical writers at $20. The newsletter ended by reminding everyone that since April 2 was John Galt's birthday it was a paid DG holiday and everyone could sleep in (John Galt is a character in Ayn Rand's 1957 novel Atlas Shrugged). For the complete story and more about DG, see Bill Foster's TeamFoster.

Along with our colleagues in other documentation departments, we worked to create a "meta-language" of embedded codes that simplified typesetting and encouraged standard conventions for formatting the text. Variants of this type of meta-language have regularly appeared in our workflows right up to today.

Metacodes for fun and profit

The embedded codes we used for our meta-language were similar to IBM's Generalized Markup Language (GML) developed in the 1960s, which evolved into the Standard Generalized Markup Language (SGML), the forerunner of the HyperText Markup Language (HTML) we use today for online content. The American National Standards Institute (ANSI) released the Standard Generalized Markup Language (SGML) in 1986, which became the basis of several subset markup languages, including HTML.

We can't overemphasize how suddenly important it was to *use computers* to create the documentation for using computers. It may be hard to imagine today, but for the most part, archived content was on paper. With computers, you could archive vast amounts of data on disks, and duplicate the disks (perhaps keeping the duplicates in a safe location), and thereby have a more secure and compact archive than you could ever have with paper. We were ecstatic about the ability to carry the contents of two books and four magazines on a few disks in a briefcase. Today, of course, such content can be stored online, where it never goes away (hopefully). You don't even have to think about it.

Back then, lots of people were thinking about it, and were determined to compete in this nascent minicomputer industry. Like the other "Massachusetts Miracle" minicomputer companies that reached their peak in the late 1970s, including Digital Equipment Corporation (DEC) and Wang Labs, Data General eventually disappeared during the massive shift to personal computing that was centered in California, victims of another generation of computer mavericks. Today Data General is best remembered because one of its computer-building projects was chronicled in Tracy Kidder's *The Soul of a New Machine*, a book that won the 1982 Pulitzer Prize.

We weren't exactly prescient in late 1979 when we decided to move to Northern California with proximity to Silicon Valley — we were ready to leave. The valley was already home to aerospace and defense industries and semiconductor companies, and IBM loomed in the south of San Jose. We had heard rumors about the congested freeways and the pocket-protector nerds at places like Control Data Corporation (CDC), and I witnessed it first-hand when CDC flew me out for an interview in a building alongside the Moffett Field airfield.

The tides were turning. At a computer convention in New York, I was blown away by the PR campaign that featured a microcomputer called the Apple II, which I saw in the back of a limousine rented by Playboy Magazine. The future had arrived.

Publishers can be predators

Berkeley, California, was the birthplace of the Free Speech Movement in the early 1960s, and a hotbed of leftist socialist politics for decades before and after. Coffee houses and esoteric bookstores proliferated in the shadow of the University of California. When we arrived, tiny businesses like Kentucky Fried Computers and North Star also were eking out a living.

When I visited an early computer book publisher in Berkeley, I was immediately charmed by a presentation by another visitor, Lee Felsenstein, who was demonstrating his microcomputer, the Processor Technology Sol. Now this was a personal computer! Housed in a friendly blue case with walnut sides that reminded me of a wood-paneled station wagon, the microcomputer and full typewriter keyboard could be connected to a black-and-white television and display enough characters on each line to approximate the width of a printed page. Rather than using simple dots or using only upper-case characters (as on the Apple

II), Lee's technology formed readable alphanumeric characters on the screen. Clearly, this was the first computer for writers, humanists, poets, and hippies.

The Processor Technology Sol. Add a display monitor to use it as a computer.

The Sol ran the first disk operating system designed specifically for microcomputers running business applications. Its inventor, Gary Kildall, called it CP/M (Control Program/Monitor, later changed to "Control Program for Microcomputers"). I recognized the command set because it resembled a minicomputer system lost to history called TOPS-10, which had also been the model for Data General's RDOS.

Kildall's company, Digital Research, captured a considerable market share in the burgeoning microcomputer industry by making CP/M available for computers from different manufacturers. More than half a million CP/M computers were sold between

1976 (when the IMSAI 8080 was introduced) and the end of 1981 (when the IBM PC was introduced).

By 1980, CP/M computers could run MicroPro International's WordStar, one of the first word-processing programs for microcomputers. Connected to a daisy-wheel or dot matrix printer, you could use the Sol to write books with WordStar and also calculate the royalties with SuperCalc, a VisiCalc lookalike for CP/M systems. (If you were born after 1990 you probably don't recognize these technologies, unless you worked in an accounting department for an old business or the government.)

The Royal LetterMaster, a budget daisy-wheel printer from the 1980s. (By Billatq - Own work, CC BY-SA 4.0, Wikimedia Commons.)

The Epson dot matrix printer (Wikimedia Commons).

Community memories

In Berkeley in 1973, Lee Felsenstein helped start the very first social network, the Community Memory Project public bulletin board system. Lee was an original nerd in work clothes, unfashionably short hair, and horn-rimmed glasses. He dropped out of UC Berkeley in 1967 during the heyday of the peace movement, and alternated between electronics jobs and work in the movement. "I had some proscription in my personality against having fun," he told Steven Levy in Levy's book *Hackers*. "I was not allowed to have fun. The fun was in my work."

Community Memory terminals were scattered around Berkeley for public use. At first it mainly provided bulletin board notices, except that responses to a notice could be within a few minutes rather than days. Soon the Berkeley hippies were entering quotations and poems into the keyboards, and information was available if you typed in something like **FIND HOSPITAL**↵.

Lee was ecstatic about bringing computers to the people. He would later call it an epiphany. He wrote in his recent book *Me and My Big Ideas*, "It finally struck me that the tool I needed to support the formation, growth and ongoing life of communities of interest would be a network of computers!" By 1980 he was designing the first portable computer, the Osborne 1, for Adam Osborne's new company.

I was salivating for the Processor Tech Sol, but it was destined for the publisher's desk. So I wrote the manuscript for *The CP/M Handbook* on my own IBM Selectric typewriter.

Of course I didn't know I was "volunteering" my first book. Manducatex (not its real name), the publisher that had hired me, was run by a pompous and litigious PhD, Dr. Roscoe Zeirocks (not his real name). His business model was to elicit manuscripts from UC Berkeley students who were readily available as unpaid interns. Dr. Zeirocks would then write unnecessary and often inscrutable introductory paragraphs for the books, in a writing style that called to mind the old *Dragnet* TV series.

So perhaps Manducatex needed a lighter touch from a good writer. I immediately agreed to write books for a pittance salary, a promise from Zeirocks (never kept) to pay moving expenses, and another promise from Zeirocks (also never kept) to give me author recognition.

We packed up and moved to California. Never mind that our first night in San Francisco our car was broken into and everything stolen. Friends said it would be impossible to find an apartment in Berkeley, but we turned our bad luck into good and found one the very next day. Never mind that the folks we met at our new favorite sunset spot were packing up to go back East because they couldn't make meaningful relationships in this hothouse of high tech. We were enjoying each other and the sunsets.

I had never before seen a book about an operating system, so I used as a reference example one of the most popular manuals of the human race, the Bible, which could be described as a combination of history and life instructions. I envisioned something like a "CP/M Bible" that would tell the story of this operating system as well as explain how to use it.

A movable feast

While the vast majority of all the books ever printed are out of print, a fair number are preserved in electronic form, and an even smaller but more significant number are continually reprinted over time, and even over centuries, demonstrating the efficacy and immutability of the print medium. In 1439 Johannes Gutenberg used a new invention at that time, movable type, to create the first printed edition of the Bible. Reprinted copies are still in use today, more than 600 years later.

When I started at Manducatex, the general manager had just been fired and five employees were getting ready to resign, including the marketing person. Zeirocks met me mid-office, on his way to somewhere else, and promised to have a half-hour talk, which never happened. Over drinks and smokes at a party of the remaining employees, who were plotting a revolt, I learned that many of them had been hired from Kelly Services as typists, that some books were authored by "The Staff of Manducatex" because the writers left town, how writers under contract were often threatened with lawsuits, and so on. They passed on rumors about his demanding wife, in leather and stiletto heels; whips and chains were mentioned more than once.

After I handed my manuscript over to Manducatex, Zeirocks promptly fired me and refused to reimburse our moving expenses. Since the type from a typewriter could be identified, the paranoid Zeirocks employed a typist using a different machine to retype the book, and then he put his name on the byline.

A lawyer counseled me to forget about it and write another book about CP/M, how many could Zeirocks sell anyway? Turns out he sold over 300,000 copies of that book. To add insult to injury, he credited me with "making editing changes for accuracy," an edit I was not allowed to do; so that any errors could be blamed on *me*.

> ### *"See if you can git it on the paper"*
>
> And let this be a lesson: Insist on a contract and pay attention to the copyright and byline paragraphs! See Warning: Your Byline May Not Be Your Byline Forever.

Nothing travels faster than bad news. A week after my firing, Gary Kildall, the inventor of CP/M who was reviewing the book for technical accuracy, contacted me about working for his company, Digital Research in Pacific Grove. By the time I drove down to that lovely town by the sea, Zeirocks had warned Kildall not to hire me and threatened a lawsuit.

Several weeks later, when I landed a job at Intel to write about the 8088 chip, Intel's formidable CEO, Andy Grove, dropped by my cubicle to ask me if I knew Roscoe Zeirocks. I nervously started to tell him my story, but he just laughed, clapped me on the back, and said "Welcome to Intel."

A Deadhead at Intel?

Around this time, out of the blue I received a phone call at my Intel desk from a stranger who asked, "Are you the Deadhead working at Intel?" Someone in the Grateful Dead community was reaching out to me — Hank Harrison, author of a history of the band called *The Dead*. This call started an intense friendship that introduced us to San Francisco's Haight-Ashbury hippie community leaders (more on that later). But in March of 1983, Harrison wrote a column for InfoWorld called Silicon Tattler, in which he described my book ripoff, referring to Roscoe Zeirocks as "R. Burlap Sacks". In Harrison's words, "Book publishing in America has a long and respectable tradition, but these traditions seem to fall by the wayside when competitive pressure builds and when opportunists and carpetbaggers jump into the field with scams designed specifically to victimize the more talented, sometimes less assertive, people." Thanks, Hank (and R.I.P. our old friend).

Back at Data General in the late 1970s, Cheryl had worked on the hardware side of the business, on the materials database. As a data entry clerk, she learned firsthand how to add to and extract the appropriate necessary information from the database to prepare a bill of materials for new products — as well as for customized computer orders. During this time Cheryl had parlayed that experience to get hired by the University of California's bureaucracy, where no innovation could occur without memos in triplicate, managing antiquated mainframe computer-bound coding sheets that were filled in by humans writing the words backward, from right to left, in order to right-justify on printed reports the descriptions accompanying UC's inventory codes.

In 1980 she quit that job to work at the Lawrence Hall of Science on the UC Berkeley campus, where she drove the "Apple Van" to elementary schools around the Bay Area, teaching children how to program Apple II computers. This was one of the first personal computer literacy projects. Eventually she became a lab assistant, teaching programming courses to Lawrence Hall of Science daily visitors who were mostly kids and college students.

The Unix man page

UC Berkeley was also the birthplace of the strongest dialect of UNIX, the operating system language developed by a group of long-haired engineers that included Ken Thompson and Dennis Ritchie. The *UNIX Programmer's Manual*, published in 1971, could be considered the first real computer manual. The documentation employed the manual page (a.k.a. "man page") format that is still in use today, offering terse reference information about command usage as well as bugs in the software, and listing the authors of programs to channel questions to them.

These experiences primed us for more computer literacy projects and books. We also helped proliferate microcomputer word-processing software for technical writers a few years before the PC arrived. Gary Kildall, who recognized the talent behind that first CP/M book, gave us multiple copies of CP/M, which had been designed to run on Intel microcomputers. I brought those copies into the Intel software department, whose engineers had never heard of Digital Research, and showed them how much better CP/M was compared to their homegrown system.

It was the start of something beautiful for the documentation department, because CP/M could run WordStar, which started many a technical writer on the road to composing on a computer rather than a typewriter. We took the lawyer's advice a year later and co-authored *Infoworld's Essential Guide to CP/M* for the nascent IDG Books, using — you guessed it — WordStar on an Intel machine running CP/M.

Tony the Deadhead from Intel (on left), with Adam Osborne, the hedonist from Osborne/McGraw-Hill and founder of Osborne Computer, in 1980

Another good turn was hooking up with Roscoe Zeirocks's arch rival, another Berkeley publisher named Adam Osborne of Osborne/McGraw-Hill Books. You could find the outspoken and self-described hedonist at all the computer trade shows, opening people's minds with a glimpse of the future of microprocessors.

Osborne was a pioneer in the computer book field, founding a company in 1972 that specialized in easy-to-read computer manuals. In 1981 he introduced the Osborne 1, designed by Lee Felsenstein, which we describe in more detail in this chapter.

Osborne directed us to hook up with Bob Albrecht, a founder of the People's Computer Company, author of *What to Do After You Hit Return* (a best-selling computer book of 1975), a contributor to *Dr. Dobbs Journal,* the first computer geek magazine, and a Saturday regular at the music scene in Fremont Park across the street from the first Peet's Coffee in Menlo Park, California.

"Time to change all that..."

The People's Computer Company, founded by Dennis Allison and Bob Albrecht, was among the first organizations to recognize and actively advocate playing as a legitimate way of learning. It recognized in personal computing a great potential for individual empowerment and social improvement. Its newsletter, first published in 1972, introduced itself: "Computers are mostly used against people instead of for people; used to control people instead of to free them; Time to change all that — we need a... People's Computer Company."

Volunteers of computer literacy

It all seemed to make sense, this intersection of the 1960s counterculture with the 1980s computer culture in the San Francisco Bay Area. One Saturday morning in Fremont Park across the street from Peet's Coffee in Menlo Park, Bob Albrecht, looking like a hyper version of Randy Newman in white button-down shirt and worn jeans, introduced us to LeRoy Finkel, respected math teacher, computer hobbyist, and 1978 Computer Using Educators (CUE) co-founder, who co-authored some of Albrecht's books. I showed him my Business BASIC manual from Data General, and he was immediately impressed. Albrecht and other partners formed an "author house" called Dymax that published books and shared the royalties. "Send us some of your book ideas," Albrecht told me in the manner of a schoolteacher handing out an assignment. "Box in the crazy ones, you know, make them stand out, and those that like sensible ideas will read the sensible ones, and those that like crazy ideas will read the crazy ones."

This park became the nucleus of our new life in Silicon Valley. I warmed up my harmonica and joined the Graceful Duck, two acoustic guitar players who had memorized every Grateful Dead song ever played. Dan Rosset, a Whole Earth warehouse employee and fan of Ken Kesey's Merry Pranksters who helped start the People's Computer Company (PCC), played his bongos. Dan was the instigator of the phrase "Give in to the grin" and the organizer of the "inspiration breaks" at PCC that involved a trip to the parking lot for a quick couple of tokes and a stiff cup of Peet's coffee. He was known to bum a ride up to Skyline Drive so he could roller skate all the way down Kings Mountain Road.

Gray-haired and bearded Harry Ely the librarian, who with his artist wife LaVerne LeRoy once provided room and board for Jerry Garcia, pounded on his hammer dulcimer. (For an interesting story about Harry and LaVerne, who always reminded Cheryl of Ruth Gordon in her artist character role in the cult classic film *Harold and Maude*, see *Psychedelic Palo Alto* by Blair Tindall.) Also on hand were Larry Tesler, the aforementioned inventor of copy/paste at Xerox PARC and at that time head of the Apple Lisa development team, and Daniel Kottke, an early Apple employee who debugged Apple II printed circuit boards and worked on the design for the Macintosh keyboard. Ed "Edrid" Riddle, one of the original Atari VCS programmers and developer of the classic game Indy 500, showed up occasionally with a guitar.

We had rented the reconverted garage of a mansion in what seemed to be a colony of aristocrats and their horses, the bucolic town of Woodside, tucked into fogbound redwoods about thirty miles south of San Francisco and a few miles west of Menlo Park and Palo Alto, connected by the financial Wall Street of Silicon Valley, Sand Hill Road. Neighbors included a slew of financial and technology wizards as well as Nolan Bushnell, Gordon Moore, Mike Markkula, Steve Jobs, Shirley Temple Black, and Joan Baez. We eventually moved up to the peak of Kings Mountain, where we could hike in beautiful redwood forest parks, and might casually encounter Neil Young dining out or shopping for a new handmade hat at the local art fair.

After I brought CP/M into the Intel software department to run on their new Intellec MDS-800 development systems, my boss trusted me enough to let me take one home to work over a weekend. The machine and its separate 8-inch floppy drive subsystem and terminal barely fit into my trunk and back seat, but I managed to set it up at home with Cheryl's approval and got productive. The experience convinced us that if only we had one of these microcomputers, we could moonlight writing books from home at nights and over weekends.

So it was that in 1980 we began our experience with working from home on a computer. My first book with my byline was the *TRS-80 Model III User's Guide* (Wiley) with LeRoy Finkel. The TRS-80 Model III was the first true "home-office computer" with its all-in-one design — the monitor, keyboard, computer, and 3.5-inch floppy disk drives were all in the same unit. By that time the novelty of the Apple II, Atari, and Amiga computers had worn off. People were disenchanted with the machines that promised so much in potential and delivered so little useful software besides games. Apple II computers were criticized in particular because the Apple II was not really a game machine like the Amiga and Atari.

The TRS-80 Model III personal computer in 1980.

The "trash-80" included Disk BASIC with text editing, and I could submit the edited manuscript to Finkel and to the publisher as a set of floppies. They could then print the drafts and transfer the text to their typesetting system.

Meanwhile, Cheryl volunteered for ComputerTown, USA!, a public access computer literacy project begun in 1979 to serve both adults and children in the Menlo Park Library. Using a grant from the National Science Foundation, the People's Computer Company developed an implementation package to assist other libraries in developing ComputerTowns, with activities including classes, workshops with hands-on access, computer rentals, club sponsorships, and game contests. The Logo programming language and turtle graphics — plus BASIC programs and games that were typed into the first Commodore PET and Apple II from the book *What To Do After You Hit Return*, then debugged and saved onto cassette tapes and floppy disks — gave many children their first experience of controlling something on a computer display. The project introduced personal computers into public libraries across the country and in the UK, as well as in recreation centers, after-school clubs, senior centers, and other non-formal learning venues. These efforts were in line with providing equal access to computers, a controversial issue in the educational community at that time.

The personal computer revolution is generally thought to have started with the spreadsheet, known back then as the killer app ("app" for application). Driving this change was the need for personal productivity and access to information locked up in the corporate mainframes managed by IT fiefdoms. The first killer app was VisiCalc, which, by itself, made the Apple II useful in business.

Before anyone knew about that, in 1980 we gave a roomful of people at the Menlo Park public library a demonstration of an early version of VisiCalc. We used an Apple II equipped with an 80-column display card and TV monitor. At first, we showed them simple, blank rows and columns and talked about how you could dream up an entire business, or at least a household budget, with these rows and columns. Reams of market research and analysis have since been published about the efficacy of spreadsheets for business and personal productivity, but at that time, no one knew what we were talking about.

The lesson we learned at this point was that we needed to do some rapid prototyping to quickly show what we were talking about. This became our first "demo" or "beta test" of our ideas for using personal computers in business. We whipped out a standard IRS Form 1040 and displayed an equivalent version of the form in VisiCalc. Never mind that the demonstration eventually crashed the system in front of everyone. If the spreadsheet could free their taxes by eliminating the accountants, we thought they would believe in the personal computer.

And so they did. Many people considered the spreadsheet to be the primary reason for getting a personal computer. VisiCalc stumbled through the 1980s and eventually lost momentum to Lotus 1-2-3, the killer app for the IBM PC. Lotus 1-2-3 lasted over a decade but was eventually dethroned by Microsoft Excel. While Excel spreadsheets are prettier, and some businesses find specialized ways to use the application and its set of formulas, there is not much more you can do with Excel that you couldn't just as easily, perhaps more easily, do with VisiCalc in the early 1980s — except, perhaps, make more complicated mistakes.

Recognizing the humor in this situation, we collaborated on a media project with the Berkeley publisher Celestial Arts which was churning out alternative lifestyle books and posters. We put together the *Murphy's Computer Law* poster with some pearls of wisdom such as:

> *Bove's Theorem: The remaining work to finish in order to reach your goal increases as the deadline approaches.*
>
> *Meskimen's Law: There's never time to do it right, but always time to do it over.*
>
> *Murphy's Fourth Law: If there is a possibility of several things going wrong, the one that will cause the most damage will be the one to go wrong.*

The Murphy's Computer Law *poster published in 1984 by Celestial Arts.*

Over the years we have learned to appreciate that this poster has been distributed all over the world, and to more people than have ever read our books. We'll continue to sprinkle a few more of Murphy's computer laws in subsequent chapters.

Faire on the mountain

Bob Albrecht at the PCC had published the entire source code for Tiny BASIC for the Altair, and had also put together a newsletter devoted to it. However, the response was so strong and immediate that Albrecht was overwhelmed, so he enlisted a fellow PCC volunteer, Jim Warren, to edit it. This became *Dr. Dobbs' Journal of Computer Calisthenics and Orthodontia (Running Light Without Overbyte)*. The dental references were jokes about saving bytes, not bites.

Bill Gates, at that time the CEO of a very small company called MicroSoft, had recently remonstrated to the Homebrew Computer Club members that copying the paper tape of his BASIC was wrong; it was *stealing*. Warren editorialized in the first issue the counter-view about the benefits of free and inexpensive software — so inexpensive that it would be cheaper to buy it than to copy the paper tape.

Free software!

How could free software, written by refugees from the Berkeley Free Speech movement of the Sixties, be taken seriously by the business world? This view was reinforced by the new shrink-wrapped model for distributing software that no longer offered any way of modifying the product. Scientists and engineers who were used to sharing source code and tinkering with software were appalled by shrink-wrapped software. Richard Stallman, founder of the Free Software Foundation and the GNU Project, wrote a manifesto in 1984 setting out the precepts of open source software and reminding everyone that programmers always used to share the source code and that sharing was essential for innovation. We took this value to heart and eventually developed another book, Free Software, about the world of public domain "open source" software.

Jim Warren had started the annual West Coast Computer Faire in San Francisco in 1977, at which the Apple II and Commodore PET made their debuts. Lines had formed around the block to get into the Civic Auditorium. Warren described the scene to Steven Levy in *Hackers*: "Nobody was pushy. We didn't know what we were doing and the exhibitors didn't know what they were doing and the attendees didn't know what was going on, but everybody was excited and congenial and undemanding and it was a tremendous turn-on."

The show would be ablaze in lights, awash with people, and as rowdy as a flea market. Warren, a large and imposing figure in blue denim, would rush by in a blur on roller skates, barking orders into an old-fashioned walkie-talkie and herding attendees into manageable lines with the unmistakable energy of the Man In Charge. He'd abruptly spin around and stop on a dime to pick up a continuing conversation with a crony, then push off again, courting the attendees with offers of help, saying "Thank ye kindly," like some kind of demented Quaker, nodding and smiling to everyone and anyone as he rolled on by.

In the early 1980s we visited his mountain hideaway on Kings Mountain in Woodside, a quasi-geodesic dome structure surrounded by redwoods with a fantastic view of the Pacific, a huge hot tub made from a 12-foot-diameter Del Monte fruit canning barrel, and an Alpha Micro minicomputer connected to a phototypesetter. He joked about how he had hired the locals right off the barstools of the Boots and Saddles Bar in La Honda to work on the Faire. Some of them were veterans of LSD guru Ken Kesey's Merry Pranksters.

Our interview ended up in that hot tub, which would be visited by some of the most important Silicon Valley journalists. For example, we shared that hot tub with Steven Levy from *Rolling Stone*, who was impressed (according to his book *Hackers*) that Warren's home also included his computerized work quarters, from which a staff of over a dozen would prepare a small empire of publications and computer shows.

Jim Warren in his roundhouse with a 360-degree view that in-cluded the Pacific (1980).

Searching libraries from the desktop

Jim Warren opened our world to searching by providing us with a Dialog account, the world's first online information retrieval system to be used globally with significant databases (see Dialog History for details). In the 1980s, a low-priced version accessed by a telephone modem (known as "dial-up access") was marketed to individual users as Knowledge Index. We learned how to phrase searches into its massive bibliographic and reference databases to get the information we wanted. At that time we were researching "videotext" systems such as the Telidon experiment in Canada.

One of these publications was *DataCast* magazine, which paired our vision of providing tutorials on how to use CP/M software with Warren's vision of "datacasting" over radio waves. This evolved into our own *User's Guide to CP/M* magazine when Warren decided to move on from publishing — he had already sold his *Intelligent Machines Journal* (IMJ) to Pat McGovern's publishing empire, which would rename IMJ to *Infoworld*. That work was an important stepping-stone in our careers from technical writing to magazine and newsletter publishing and multimedia production.

Documentation in a magazine format

As we entered into book contracts in the 1980s, we were still bound up in a wasteful distribution system that favored the largest publishers at the expense of writers. The book distribution channel, like most distribution channels that rely on warehouses, trucking, and retail outlets, was dominated by the "distribution mafia" that siphoned off more than 50 percent of the revenues. A set of large publishing companies dominated the bookstore racks, and deals with those publishers were not attractive, nor did we feel that they paid enough attention to quality.

Driven by the need to create a better product devoid of typesetting and page layout errors, we decided to control the entire production process and publish our content not as books but as a magazine to which readers could directly subscribe. The magazine's "letters to the editor" section would engender reader feedback that would include corrections, opinions, and new developments. The content, over time, would form a documentation library for the computer user.

Magazines about computers, such as *Creative Computing* and *BYTE*, were popular among professionals in the computing industry of the early 1980s. The latter evolved from an amateur radio magazine called *73*, published by Wayne Green. Jim Warren had a long-standing feud with Green over magazine ethics, going so far as to publish a "Wayne Watch" column in the *Journal* about Green's bitter divorce and his IRS problems; Green's ex-wife later sold *BYTE* to McGraw-Hill.

Creative Computing was more to our liking. David Ahl, a former engineer at DEC, wanted to put out an educational magazine about computing in the mid-1970s. He used his own funds to print 11,000 copies of a flier that he sent to Hewlett-Packard and other minicomputer vendors, which resulted in 850 subscriptions to a magazine that did not even exist yet. Instead of printing 850 copies, Ahl split the subscription money in two; he kept one half for future operations, and used the other half to print around 8,000 copies. He sent copies to the subscribers first, and sent the rest for free to a wide variety of companies, libraries, and schools. Weathering the storms of divorce, embezzlement of $100,000 in advertising revenue, and Ted Nelson as editor (briefly), Ahl eventually sold the magazine to Ziff-Davis.

Our first public collaboration (with the byline "Tony Bove and Cheryl Rhodes") was a cover story for the People's Computer Company (PCC) magazine called *Recreational Computing*, cast in the same mold as *Creative Computing*. Bob Albrecht at the PCC had introduced us to the editor, Marlin Ouverson, and he published "Sesame Place: Learning, Playing, and Using Computers" in the May-June, 1981 issue. At that time Sesame Place, an educational theme park near Philadelphia affiliated with Sesame Street, offered the largest number of computers ever to the public for access on a daily basis (55 Apple IIs with over 40 games linked by a network).

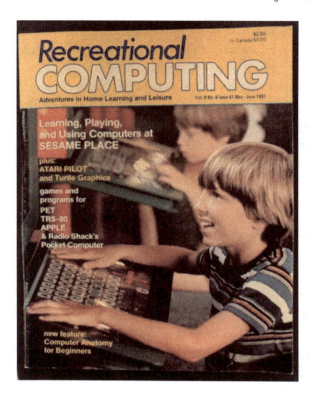

Recreational Computing *from the People's Computer Company (PCC).*

The magazine gave us a chance to advocate the use of educational games for computer literacy projects, using Sesame Place's unconventional approach as a prime example. "What makes Sesame Place so unusual is that it combines a physical playground with a conceptual playground," said Dr. Arthur Luehrmann, computer research director at the Lawrence Hall of Science at that time who had coined the term computer literacy. "At Sesame Place young people can learn on their own by interacting with the science experiments and computers — hands-on experiences that lead to discovery learning."

The new "computer magazines" gradually de-emphasized the do-it-yourself electronics and software articles to focus on product reviews, which drove advertising sales. Readers would have to wade through pages of ads to find useful content, and usually, the content was about a machine or piece of software the reader didn't have.

Our idea was to publish a system-specific magazine that provided tutorials on how to use that system and the software that ran on it. While dreaming of a wide-circulation magazine like *MAD*, which carried no advertising, we put together a magazine of useful content with a sense of humor.

It seems quaint today in the era of internet browsers and Microsoft Windows, but at that time it was a novel concept to use the same operating system on different computers, and the practice brought relative stability to what was once the "hobbyist" personal computer world. We ran a business using CP/M on computers as diverse as the Compupro multiuser system, the portable Osborne and Kaypro machines, the desktop Zenith, Alspa, and Xerox computers, and Lee Felsenstein's ground-breaking Processor Technology Sol. We even installed CP/M on an Apple II using a CP/M card (ironically from rival Microsoft). So it was possible to write articles and tutorials about software, such as Turbo Pascal or SuperCalc, that ran on top of CP/M on different computers.

Bringing it all back home

Gary Kildall, founder of Digital Research and inventor of CP/M, saw the wisdom of our approach and granted us free access to the Digital Research customer list of 50,000 CP/M users. In one mailer to entice subscriptions, using a public-domain image of a Native American on a horse in the desert with his hands in the air, pleading the Almighty for a drop of water, we achieved a *six percent response*, which was 100 times greater than a typical direct-mail campaign.

Looking for information about CP/M? This was the first promotional campaign by User's Guide to CP/M.

Results in hand, we looked for funding. Eugene, a good friend from my college days with a drug habit, needed a place to park some money before he snorted it all. His father agreed that we should use $20,000 to start our fledgling magazine. Jim Warren offered us the Datacast subscription list, but we were already heading off a financial cliff due to an excessive number of issues returned by the USPS for incorrect addresses and the return-postage fees.

Then we literally ran off a cliff on slick Skyline Drive in the rain, while heading into the Computer Faire in San Francisco to promote our new magazine. The car was totaled and its cassette deck was stolen by the towing company. A few stitches later we arrived at the Faire to learn that we had sold out all the issues on hand. *User's Guide to CP/M* magazine was launched.

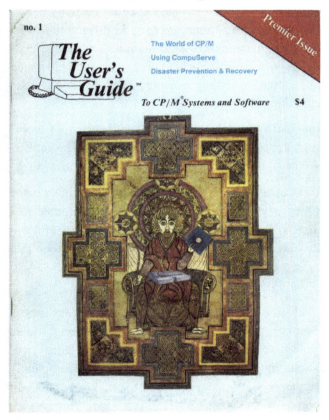

First issue of User's Guide to CP/M *in 1982 with an image taken from the* Book of Kells *and doctored to include a keyboard and floppy disk.*

The first issue's cover, suggested by Hank Harrison, was an enhanced version of an apostolic page from the *Book of Kells*, an

Irish illuminated manuscript of the 9th Century, depicting John, the author of the book of Revelations in the New Testament of the Bible. Like the Lindisfarne Gospels and many other Dark Age illuminated manuscripts, the *Book of Kells* is loaded with submerged codes. The keys to deciphering them are found in the front of the book in a section called the Canon (much like a computer program heading or man page). This is usually a section that enables the reader to read the texts in differential order to arrive at different meanings or clarification. This same technique was carried out in the Kabbalah and in many pre-Christian Gnostic texts. In this sense they operated very much like user guides to culture and religion.

The Whole Earth Catalog with its byline "Access to tools" had been an inspiration for doing the *User's Guide*. The first issue acted as canon to unlock the mysteries of CP/M.

At that time every "mass-market" personal computer used CP/M, as well as nearly every business desktop computer system, largely due to its familiar set of commands and useful application programs. We considered it to be a Rosetta Stone that could unlock the potential of the computer.

The operating system was a medium for software exchange that enabled us to use different programming languages and run a mix of proven programs from older and larger systems. It also became a medium for data exchange — you could edit documents in files that you could transfer to other, almost completely different computer systems, and still use them. We would edit articles electronically and transmit them to a typesetting service using CompuServe, the first major commercial online service in the world, without ever printing the articles on paper.

A prediction from 1981

From our first issue: *CompuServe is a service you can call by phone and use with your home or office computer. Such information services will soon provide electronic banking, "tele-shopping," reservations by computer, research databases, etc. They may revolutionize the way we bank, shop, plan trips and do business.*

A look at some of the instructions gives you a feeling for how important a role the documentation played in this activity. You would never know how to copy files from one floppy disk to another without these instructions:

Using Two Disk Drives

The following procedures assume that you have two floppy disk drives: drive A and drive B. If you use a hard disk, skip to the "Hard Disk" instructions.

1. Leave your system disk in drive A.

2. Insert the receiving disk (the one to hold backup copies) into drive B. Be sure that the receiving disk is not write-protected.

3. If your receiving disk is new (never before used with CP/M), it probably needs to be formatted (look back at "Formatting Disks").

4. What about the original (source) disk to be copied? Put it aside for a moment.

5. Type the command **B:** and press your RETURN or ENTER key (shown as ↵).

6. When you see the B) prompt, hold down the CONTROL key and type a **C**. The B) prompt should reappear.

7. Type the command **A:** (followed by ↵), wait for the A) to appear, and then type the command **PIP** (followed by ↵). An asterisk (*) should appear on a line by itself.

8. Take your system disk out of drive A and replace it with the original (source) disk to be copied.

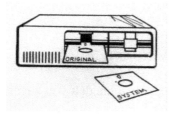

9. Type the following command (followed by ↵):

    * **B:=*.*** ↵

10. Watch the names of the files go by. They are all being copied from drive A to drive B. Be patient.

11. When it finishes, take your original disk out of drive A and your receiving disk out of drive B, and put the system disk back in drive A. Put a label on the receiving disk to show that it is a backup of the disk in drive A.

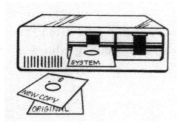

Note: You should never write directly onto the floppy disk, because too much pressure from a ballpoint pen or pencil will destroy the data. Write on the label before attaching the label to the disk.

Without the above instructions, you would have been stuck. This explains why we received so many letters to the editor like the following two letters:

Congratulations on a great start with the User's Guide. It's very well done and way beyond the quality of a newsletter. I hope with all of the CP/M users out there that you get all of the support (subscriptions) and all of the input (content) that you'll ever need to keep it going on...

I am impressed. Having read a lot of magazines and books concerning computers and operating systems, I have found few that really provide the reader with a sense of profit in his investment — that profit being a meaningful learning situation...

The tiny publishing staff, working out of a post-office box at Stanford and a rented cottage at the top of nearby King's Mountain in Woodside, included cartoonist Lon Shoemaker, his brother Bill Shoemaker from the Menlo Park Post Office, photographer Paul Winternitz, Sahnta Pannutti (the production designer from InfoWorld), the previously-mentioned Ed Riddle, and layout artists Chrisann Brennan and Tina Redse, both personal friends of Steve Jobs. Contributing editors and writers included notables such as Steve Rosenthal, Kelly Smith, John Barry, Dana Blankenhorn, Arthur Naiman, Paul Saffo, and Jonathan Sachs.

If you had visited us in the Fall of 1982, you would have found half our kitchen taken up with two Alspa ("bookshelf") CP/M computers with two external hard disk drives connected by SCSI Parallel interface cables. Each computer had its own ADDS Viewpoint terminal with a keyboard. We also used a portable Osborne 1 on a card table.

The Osborne 1 portable computer (Wikimedia Commons).

The Hayes Smartmodem lights blinked as we downloaded the latest draft of an article from a contributor. On one display was a regular draft, and on the other was a version of the draft with embedded metacodes standing in for the real typesetting codes.

Over in the corner, a Diablo 630 daisy wheel printer with a military-specification power supply the size of a file cabinet churned out our latest promotional letter to subscribers. An Epson MX100 pin-feed dot matrix printer in the other corner spit out mailing labels.

The Osborne 1 received the files ready for typesetting, and we would then lug the Osborne over to George Graphics in San Francisco to get the typesetting done.

The personal computing renaissance

Success with our magazine was measured by the number of subscribers, not "newsstand" or retail sales, which is always minuscule for a niche readership. We could be proud that the magazine was circulated across the land through book-stores and retail outlets as diverse as The Com-puter Store in New York City and City Lights Bookshop in San Francisco, but sales didn't translate into much revenue. Nevertheless, we designed the magazine's cover to "pop" among the racks of magazines preaching about the per-sonal computer. We used classical art modified to be a pun on high technology. For example, the second issue's cover art is Michelangelo's Isa-iah, with additional features by artist Sahnta Pannutti. "The prophet may have just noticed the computer as a research and writing tool, and may soon announce that the personal computing renaissance is upon us all."

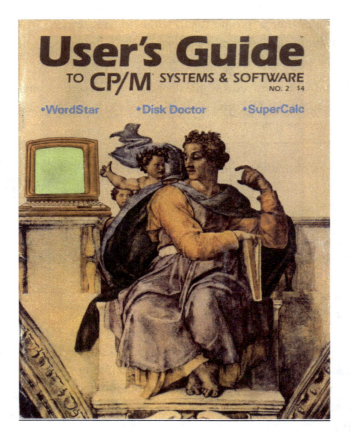

Look, Dad, you can use a computer to write your next book!
User's Guide *No. 2 (1982) with artwork added by Sahnta Pannutti.*

Compared to today's content sites, retail pricing and subscription pricing for magazines were complicated affairs governed by the United States Postal Service (USPS). For example, the single-copy price was $4.00 (or $5.50 if sent first-class). A subscription for one year was, in the U.S., $18 if payment accompanied the order, or $21 if the subscription was to be billed.

Add $10 for first-class delivery. For foreign subscriptions, the price was $24 U.S. for one year (surface mail) if payment accompanied the order, or $29 U.S. if the subscription was billed, and you had to inquire about rates for airmail delivery. Good luck with that!

Fortunately, many subscribers requested to buy all the issues once they read a single issue. Many back issues and subscriptions were sold, ensuring a steady income stream from our small business to the USPS.

The first days are the hardest days

Failure was inevitable for our *User's Guide* because we had no sales talent to make our case to potential advertisers. We had applied the newest technology to magazine production and distribution, with desktop computers for editorial content and multiuser computers for organizing databases and preparing mailing labels. We had used the newest spreadsheet software to create a unique magazine publishing model for a "hobbyist" magazine based on newsstand circulation. And we tried, and failed, to sell ads. Eventually we had hired a media organization in Los Angeles that was focused on the new computer magazines. They couldn't sell ads either.

The price of an ad depends on the magazine's circulation; if you can grow the latter, you can reap higher prices per ad. However, growing the circulation of a magazine is the most expensive aspect of the business, just like growing the traffic to a website, blog, or fan page. With printed publications, a huge expense was paper not only for the magazine but for direct-mailers, and one of the chief discussions was arguing whether or not to buy a train car of paper at a time for a cheaper price.

Computer show booths were costly, but occasionally we exhibited in order to show the market that we existed. The retail take at such shows was minuscule. Once while leaving the Faire, we dropped our coin box on the sidewalk outside the show, and media mogul Patrick McGovern, perhaps the richest publisher on the planet at that time, stooped down with us to pick up the rolling coins.

Another huge expense was production, which involved numerous people cutting scrolls of typesetting and pasting them onto pasteboards, which were then photographed by plate-making machines.

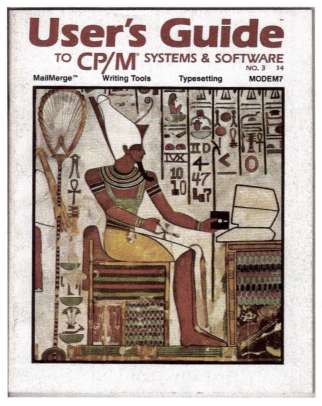

User's Guide *No. 3. In 1983 we shared our experiences with writing and typesetting the magazine's content with CP/M computers. Artwork added by Sahnta Pannutti.*

A fond memory was convincing the largest typesetting firm in San Francisco, George Graphics, that we could directly pipe our content into their computer-based typesetting system. The first portable computer, the Osborne 1, designed by our friend Lee Felsenstein for Adam Osborne's new computer company, looked and, at 23 pounds weighed about the same as a portable sewing machine. We lifted it onto a nearby desk, connected an RS-232 serial cable to it, stuffed our mini-floppy disks into the two drives, and as half the company looked on to their amusement, transferred our text laden with typesetting codes into their process, thus requiring no further proofreading on our part. We received long scrolls of typeset text, and then set about cutting (there was no copying yet) and pasting onto boards to deliver to the printer to be photographed to produce the printing plates used by the Heidelberg printing press.

The guy on the left doesn't stand a chance.

The Osborne 1, at 23 pounds, weighed about the same as a portable sewing machine.

During this time period we used desktop and portable CP/M computers to produce manuals, documents, spreadsheets, books, and computer magazines. For educational software and VisiCalc for spreadsheets, we used an Apple II computer. We also used a

CompuPro CP/M system to manage our mailing list and print mailing labels. We used a "letter quality" daisy-wheel printer (such as the Diablo HyType) for correspondence and typewriter-style output, and the new dot matrix printers (such as the popular Epson MX-80) for rough printouts of gray images. We saved up for an Epson MX-100 to print our subscriber mailing labels, and paid a service to paste the mailing labels onto the subscriber copies and mail the issues.

User's Guide *No. 4 (1983). We shared our experiences with using dBase and SuperCalc to manage our publishing business. Artwork added by Sahnta Pannutti. Can you spot the computer in the image?*

The most popular software packages for CP/M included WordStar for word processing, MailMerge for form letters and printing from mailing lists, SuperCalc for calculating on spreadsheets, and both CBASIC and MBASIC for BASIC programming. We also used dBase for databases, and BASIC, Pascal, Lisp, and Smalltalk for programming simulations and examples for manuals and books. We learned phototypesetting system encoding, which we generalized into a "meta" language for typesetting. We also used a new system called TeX for typesetting formulas.

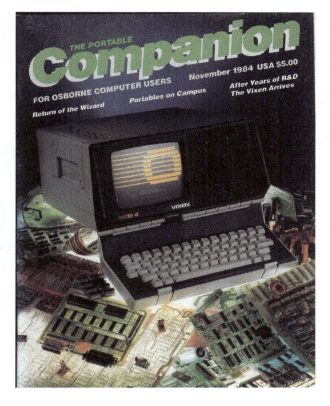

The ground-breaking Vixen portable computer from Osborne Computer Corp. ran CP/M; it was put back in the closet when Microsoft introduced DOS and Osborne went bankrupt. Cover by Paul Winternitz for our Portable Companion *magazine in 1984.*

Our first instructional video

We experimented with video tutorials decades before YouTube. Our first video centered on using the Osborne 1 in remote locations. We shot it on Mt. Diablo using KTVU Channel 2's corporate ad production team. An actor used the Osborne on the hood of a Jeep while a fantasy fire raged nearby. We then shot the actor using it with an acoustic coupler in a telephone booth (remember telephone booths?). The Osborne had a five-inch screen, a full-size alphanumeric typewriter-style keyboard, two floppy disk drives, and several ports for connecting a modem and printer, in a case the size of a portable sewing machine that weighed under 25 pounds and could fit under an airline seat.

CP/M included a standard eight-inch single-density single-sided floppy disk format known as IBM 3740 because it was invented in 1973 for the IBM 3740 Data Entry System. If you maintained data files on these disks, you could use them on other CP/M computers that offered eight-inch floppy drives, and nearly all of them did. This kind of universal compatibility was new to the computer industry.

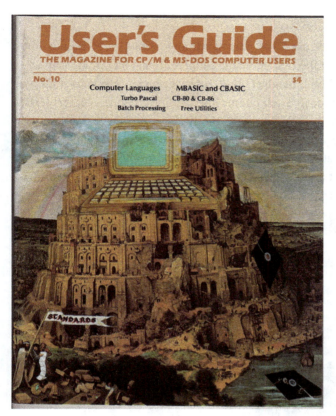

Making progress with standards in 1984. User's Guide *No. 10 featured a towering tribute to Pieter Bruegel. Artwork added by Sahnta Pannutti.*

Getting sloppy with floppies

The 5.25-inch "mini floppy" disk became ubiquitous as Apple developed a drive for the Apple II. By 1980, the 5.25-inch floppy grew in popularity but its format was proprietary for each type of computer, including CP/M computers. See History of the floppy disk.

Modems, used for data communications over phone lines, came in two flavors: acoustic couplers that you could connect to your telephone handset, and direct-connect modems that you could connect to a telephone line. Bulletin board systems (BBS) and XModem software were the lifeblood of the CP/M community in the early 1980s. The BBS was the true computer-based communication medium for the general population, long before the internet and its ancestors.

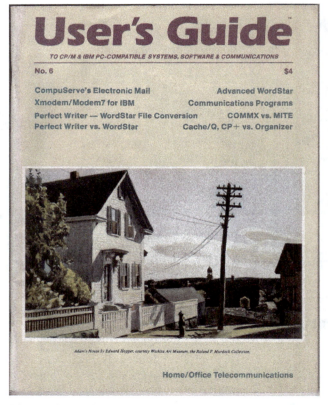

User's Guide No. 6 (1984) featured an original Edward Hopper work of art, Adam's House, to illustrate the coming telecommunications age. We also added coverage of the new IBM PC compatible computers.

A BBS is a computer that you can dial into using a modem to send and retrieve information, software, and messages. The computer runs software that allows users to connect to the system using a terminal program, and essentially run the computer, putting the computer itself in the primary role of a medium for a network, just like the social networks that evolved in later years. You could upload and download software and data as well as exchange messages. In the early 1980s the BBS systems grew in popularity and functionality, offering direct chatting and email. Some offered online games in which you could compete with others, all connected through modems.

As a system for computerized communication, the BBS evolved from the Community Memory project in Berkeley (described earlier). Early boards were crude, buggy affairs run by individuals known as *sysops* (system operators) out of their homes as a hobby. Many sysops wrote their own software, but CP/M offered the opportunity to create a de facto standard. Hackers Ward Christensen and Randy Seuss created the first, and by 1985 there were over 4,000 BBS systems in the U.S. you could dial into. By 1990 there were about 30,000 boards, covering every topic imaginable. Some were part of the "underground" — according to Bruce Sterling in *The Hacker Crackdown*: "… hackers and phone phreaks, those utter devotees of computers and phones, live by boards. They swarm by boards. They are bred by boards."

Imagine our surprise when our first editorial contributor, Kelly Smith, contacted us in 1981 about our fledgling magazine through a message on a BBS. We used a BBS system extensively, not only to exchange messages and download software, but also to receive drafts of articles for our magazine and send them out for technical review. It would take over an hour to send or receive a single article, so we had to add another home phone line just for "networking". The BBS became the hub of our editorial activities with contributors and software vendors.

But then, Microsoft rolled out DOS, a virtual copy of CP/M, along with the introduction of the IBM PC. The PC clone industry was born, and nearly every manufacturer of CP/M machines went out of business. For us, it meant that all our CP/M-based applications were useless, our third-party add-in cards for the standard CP/M computers were obsolete, and our documents and databases were locked on disks that would never be read by the new wave of PCs.

Welcome to the new machine!

Tools for print

The following are tools we used for the print medium.

Media: Text, manuals, documents, spreadsheets, books, computer magazines

Tools and languages:

CP/M, WordStar, VisiCalc, T/Maker, Xmodem, BBS, TeX (typesetting)

BASIC, Pascal, Lisp, Smalltalk

Dialog (search)

Prepress photographic processes and plates, offset printing, reams of paper, buckets of ink

Publishing from the Desktop

Freedom of the press is limited to those who own one. — A. J. Liebling

Please leave a quarter for beer. — Squid Ink marketing campaign

Franklin's Rule: Blessed is the end user who expects nothing, for he/she will not be disappointed. — Murphy's Computer Law

In 1975, I took out my righteous worldly indignation on the hapless Tufts University English Department professors about their restrictive, slickly produced literary magazine that catered to furthering the careers of professors. I published an alternative "underground" literary magazine that catered to getting students published. Loyal to the cause of student rights, we didn't care about capital expenditures or marketing; we simply sold copies for $0.25 each, and used the money to buy beer.

Today you would probably take out your righteous worldly indignation in Facebook posts, X (formerly Twitter) tweets, YouTube videos, blog entries, webpages, e-books, and so on. You can publish your thoughts essentially for free and make them available to the world. You just can't get the world to notice those thoughts and respond with cash.

In the late 1970s and early 1980s, the economics and mechanics of publishing books, magazines, newspapers, and newsletters required large capital expenditures for typesetting, page layout, printing on paper, distribution, retail sales, and direct-mail marketing. Four-color copper plates were used to print the magazines, books, posters, marketing materials, packaging, and direct mailers that drove the global printing and publishing industry.

In late 1979, a group of Apple Computer engineers and executives led by Steve Jobs visited Xerox PARC to learn about the mouse, windows, icons, and other new user interface technologies that were all the buzz in Silicon Valley. At that time you needed about $100,000 for equipment to play the typesetting game. A priesthood had developed around the technology. As an outsider to both the publishing circles and the typesetting priesthood, Jobs could see clearly that the combination of a graphical desktop computer and typesetting would be a major factor in changing the economics of publishing and allow people like us, as entrepreneurs, to produce books and magazines. By 1984 Jobs introduced the Macintosh, which ultimately changed everything in the publishing and computing industries.

The Apple Macintosh made desktop publishing possible. Photo courtesy of Apple Inc.

Some of the economics and mechanics changed with the introduction of desktop publishing tools such as PageMaker and Adobe Illustrator in 1986. But despite the rhetoric of revolution that accompanied the desktop publishing phenomenon, there wasn't as much of a revolution in publishing in the 1980s as there was in the 1990s. As book authors, magazine publishers, and newsletter editors, we still had to deal with printing, distributing the printed material, selling it at the retail level, and mailing printed promotional material.

Much of what happened in that decade was the wresting of typesetting and page layout controls from the priesthood within the graphic design industry, which *was* revolutionary. What also occurred was the development of skills that would give people the power to publish electronically and enhance their pages with graphics, images, and text typeset with professionally designed fonts.

The organizing principle for this evolution was competition over standards. Microsoft bulldozed its way into the desktop publishing industry with Word for word processing, and threatened to produce a better page makeup tool with its proprietary file format, spurring rivals to innovate in order to compete. Microsoft and its rivals cooperated to create standards for transferring graphics and images from one tool to another, but could not standardize proprietary fonts within the different computer systems — when you laid out pages on one system and then transferred the pages to a different system, different fonts were substituted, with unpredictable results. As you learn in this chapter, a "font war" developed between Microsoft (allied with Apple) and Adobe Systems that captured everyone's attention and spurred even more innovation.

Less than a decade later, the changes wrought by desktop publishing transformed the entire printing and publishing industry from a toxic, dangerous chemical and mechanical process to a fully electronic system. These changes also paved the way for information, in the form of graphics, images, and text in fonts, to be distributed and consumed digitally through display screens and internet connections.

"You say you want a revolution..."

This revolution was more than just on paper. Desktop publishing began when the cost of laser printers dropped and software became available to offer near-typeset-quality text and graphics printing. More importantly, the inexpensive publishing tools made it easier for personal computer users to do publishing production tasks without resorting to typesetting services and graphic design houses. These tools and the skills to use them were eventually carried over into online publishing (minus the paper).

We didn't consider desktop publishing tools useful at first for commercial and corporate publishing — not until they were made compatible with typesetters, film recorders, and printing-plate machines. The industry-wide acceptance of a common language of typesetting, called PostScript (developed by Adobe Systems), made it possible for desktop publishing software to produce typeset-quality text and high-quality photographs. Steve Jobs liked PostScript enough to invest in Adobe and make Apple's first LaserWriter a PostScript-compatible machine.

Of course there was resistance. Experts in any field rarely want people to understand what they do. At that time many professionals in the traditional book and magazine publishing industries resented the idea of hands-on working with personal computers. Poorly designed results from desktop publishing seemed to prove that publishing should be left to the experts who have design skills.

An editor from a large book publishing company once explained why she was adamantly against allowing writers — even those who wrote instructional books — to make copy-editing changes electronically. She saw writers as *adversaries* in the publishing process, always trying to assert themselves in editorial decisions and for the most part unwilling to make the changes. And yet, most writers consider themselves partners in this process, not adversaries — unless accuracy is endangered. Mistakes in judgment on style or grammar, made by someone who is not knowledgeable on the subject, could be much more disastrous, especially in technical books and manuals. A seemingly superficial change made by a copy editor without consulting the author could change the entire meaning of the sentence.

The resistance from the typesetting and design industries turned out to be beneficial. Professional typesetters and font designers pushed quality improvements onto the PostScript implementations that drove laser printers and typesetters and the page layout programs that offered typesetting features. More fonts were developed, kerning became more exact, and color images were handled with a newfound ease.

From the mid-1980s onward, if your company produced printed sales literature, marketing brochures, flyers, newsletters, advertisements, operating manuals, or other business communications, you could save time and money, and maintain control over the production process, by using a desktop publishing system.

Such a system, dominated equally by the Apple LaserWriter and Macintosh, PostScript, and Aldus PageMaker, put us in business as magazine and newsletter publishers, and spawned hundreds of thousands of niche-market publications and 'zines. The ability to create page layouts made possible graphic novels and comics for international readers. The page layout programs laid the foundation for the web pages we take for granted today.

By 1988, neon "Desktop Publishing" signs glowed in storefronts like Copy Plus and Kinko's, as anyone with a couple of Macs and a LaserWriter could start a service bureau. A profound moment during this "revolution" occurred at a publishing conference in 1988 when Paul Brainerd, chief of

Aldus and developer of PageMaker, held up a book that Cheryl and I had co-authored, *Desktop Publishing With PageMaker*, translated into Polish and printed on a LaserWriter. He proudly explained that the combination had been instrumental in producing broadsheets that helped the Solidarity movement win the Polish elections. Page makeup programs had indeed transformed the world by giving publishing power to the people.

Pages "en Regalia"

The catalyst for desktop publishing was the combination of page makeup software and a page description language. The Mac provided a key ingredient, and so did the Apple LaserWriter, but the glue that held them together was the page makeup ecosystem that developed around the PostScript page description language developed by Adobe Systems.

Desktop publishing would not have happened without PostScript, which included scalable typefaces and smoothly-drawn graphics. For example, the Hewlett-Packard LaserJet was very popular, but as John Warnock, co-founder of Adobe, described it in 1985, "It goes nowhere. It's a daisy wheel replacement. It has no growth potential in type, no growth potential in graphics, no growth potential in communications." (See Antonoff: How Steve Jobs helped launch the desktop publishing revolution, Money column, special to USA TODAY, Feb. 12, 2015.)

The combination of the Macintosh, PageMaker, and Laser-Writer marked the beginning of the desktop publishing revolution. PageMaker helped popularize the Mac platform and eventually led credence to Microsoft Windows as a desktop publishing platform.

A page makeup program like PageMaker played a central role in a desktop publishing system — and was a precursor to today's web publishing tools in which text, graphics, and photo elements are put in and pages come out. We used a word processing program and a graphics program to separately create the text and graphics, and PageMaker to bring them together on a page. The page makeup program let us adjust the page design at the same time that we were placing the elements. Gone were the cut marks of sharp paste-up knives, the FPO ("for placement only") notes, the heady (but toxic) smell of rubber cement paste-up glue, or the noxious odor of hot melted wax.

PageMaker simulated the procedures that artists and designers used in conventional page makeup, but all the work occurred on the screen. You could duplicate page elements and move blocks of text around the graphics, even wrap text around irregular shapes. Columns of text could be linked so that changes made to a column caused a ripple of reformatting to occur automatically, without losing lines of text at the tops and bottoms of pages. You could even "pour" text onto multiple pages quickly to see how many pages the text would fill. Designers and graphic artists had no trouble understanding how PageMaker worked; the precision of its rulers and various page displays made it very easy to line things up without the need for a T-square and a fluorescent light table.

PageMaker was not the only page makeup program. QuarkXPress on the Mac captured the higher end of the publishing market, especially magazines, with capabilities for fine-tuning complex page layouts. First out of the gate, before PageMaker, MacPublisher set the standard for "what you see is what you get" (WYSIWYG) page makeup. Ready, Set, Go! followed soon after, offering a grid approach to page makeup.

PageMaker utilized the page metaphor to give you a WYSIWYG view.

Whatcha See Is Whatcha Get

The page metaphor dominated the desktop publishing industry, and the page layout programs laid the foundation for the web pages we take for granted today. It popularized what was then called the WYSIWYG (pronounced "whiz-ee-wig") view. Thanks to The Dramatics for "Whatcha See Is Whatcha Get," released in 1971. Honorable mention goes to Tina Turner for her follow-on, "What You Get is What You See," released in 1986.

For Microsoft DOS, Ventura Publisher (which became Corel Ventura in 1993), released in 1986, was a strong contender for lengthy documents and books. At that time Microsoft Windows was still a buggy product that no one used, so Ventura Software relied on the GEM extension to provide a graphical user interface necessary to show the page metaphor.

With the page metaphor in action on the computer screen, designers, artists, and graphics-oriented people could see the power of using the computer for preparing pages for printing. As color photographs and graphics were integrated into page make-up programs, you could even prepare four-color separations directly from the computer. The tools were all digital, but the medium was still in print, on paper.

Desktop Publishing magazine

The medium became our message in the Fall of 1985 when we started *Desktop Publishing* magazine, with the vision of using personal computer publishing tools to put out a magazine about personal computer publishing tools. It was completely written, edited, and produced on personal computers. As far as we know, it was the first magazine to be typeset in this manner, using Macs, PageMaker, and a PostScript typesetting system.

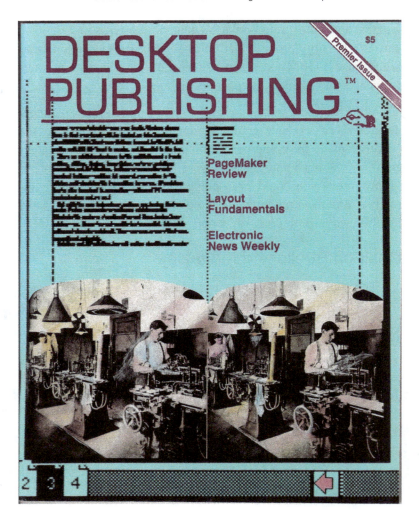

The first issue of Desktop Publishing *magazine introduced PageMaker and other Mac-based tools, with a cover designed by Paul Winternitz.*

David Bunnell of PCW Communications (*PC World* and *Macworld* magazines) shared our vision of a desktop-published magazine, and bought our publication in the Spring of 1986, renaming it *Publish!* magazine. The monthly became a publication of International Data Group (IDG), the world's leading publisher of computer-related information, and expanded to cover all forms of electronic publishing and distribution, focusing not only on high-end print production but also on CD-ROM, Web, and other nontraditional publishing media. The magazine eventually changed its name to *Publish RGB* and became a website in 1997.

Our story is not much different than the stories of software startups in Silicon Valley in the 1980s, simultaneously flush with encouraging press reports, and yet strapped for cash. By then we were running out of steam with the *User's Guide*. Stewart Brand's Point Foundation had declined to take it over and fulfill subscriptions. The choice was to fulfill the subscriptions with a PC-oriented version that covered MS-DOS, or with an entirely new magazine. In the last issue (#17) of *User's Guide* we reported on our new entry, *Desktop Publishing*, and confessed that we were not getting enough sleep. "We wear more hats around the office than the Madhatter of San Francisco's Schlock Shop."

We had gone to Apple begging for support in the form of a full-page ad in our first issue, preferably on the back cover or inside front cover. As it turned out, Hewlett-Packard bought the back cover with an ad for the LaserJet. But we were determined to get ads from the "A" list of desktop publishing — Adobe, Aldus, and Apple.

Adobe agreed to advertise, likely because the first issue would feature an interview by August Mohr with John Warnock, co-founder of Adobe. In September of 1985, we were in Adobe's lab and Warnock himself helped us pull the pages of our first issue from the first PostScript-enabled Linotronic imagesetter.

Paul Brainerd, founder of Aldus and developer of PageMaker, was at first concerned about the magazine's title. "We are using 'desktop publishing' for our product line," he told us. We pointed out that Apple was already using the term with a trademark in a magazine ad, and that we had already trademarked the term as a title for a magazine. In those days you could apply for a trademark for a magazine title by promoting and then accepting a subscription from someone in another state. After using it for a year or more, you could register the trademark. We had done this, but we were willing to let Aldus use the term for its software without a trademark. As a result, Brainerd agreed to support our effort and advertise in our magazine.

Brainerd also helped us get Apple to stop using the term as a trademark. We gave Apple a free full-page ad in a gesture of support.

The second issue of Desktop Publishing *magazine covered PC publishing software and featured a cover designed by Paul Winternitz.*

Other adventurous publishers were moving into self-publishing; most notably Michael Gosney with *Verbum*, the first art magazine to focus on interactive art and computer graphics. The self-proclaimed "journal of personal computer aesthetics" placed more emphasis on creativity than on technology. Its goal was to serve the desktop publishing industry — vendors, consultants, and the relatively limited ranks of publishers — with inspiring examples of what could be done with desktop publishing technology.

At the same time the computer magazine business was growing exponentially. *PC* magazine and its competitor, *PC World*, were both the size of telephone books, chock full of ads you had to wade through to find useful content (usually 60-65% ads and 35-40% editorial). While hobby-oriented magazines, and a few unique ones like *MAD*, could be run by enthusiastic writers, professional magazines need to be run by publishers who prioritize profit above everything else.

As CP/M companies went out of business or adapted to selling PC-compatible software, the business for the *User's Guide* almost completely unraveled, threatening to derail *Desktop Publishing*. Creditors nipped at our heels, former employees shrugged and disappeared, and our newly hired gung-ho ad sales rep turned out to be a rip-off artist. Equipment started walking out the door. We had to change the locks suddenly during the last month of business. The atmosphere was not conducive to a rebirth. We all wanted out.

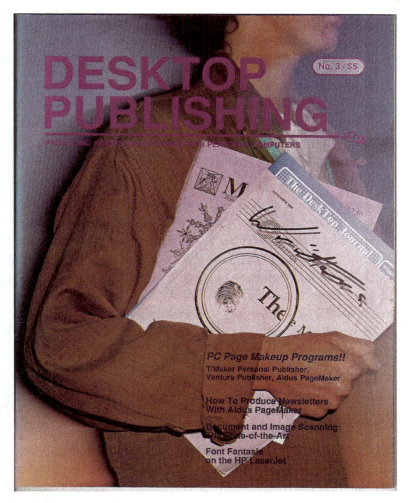

The third issue of Desktop Publishing *magazine covered PC page makeup software such as Ventura Publisher and how to produce newsletters with Aldus PageMaker. Cover designed by Paul Winternitz.*

And then a call from an advertiser jolted us. Was this San Francisco address on the invoice correct? We then realized why Shaun, the ad sales rep, had borrowed a laser printer from the office. As with the case of *Creative Computing* a few years earlier, the ad sales rep embezzled money by diverting advertising invoice payments. The new desktop publishing tools enabled his nefarious scheme that involved reproducing our letterhead and invoices with his address.

I called Paul, our photographer and cover artist, and we converged on Shaun's apartment in San Francisco. It took all of two seconds for Shaun to plead his apologies and hand over the checks, printed material, and laser printer. But that wasn't all. During a job interview, Shaun provided the advertising manager of *PC World* with a copy of the subscriber list.

At some point in the month between our first meeting with Bunnell about acquiring the magazine and the incident with Shaun, the world was coming to an end: Bunnell had decided to walk out of the deal. His staff had balked at the idea that the magazine would be produced by its editors, using primitive desktop publishing technology. *PC World* and *Macworld* used conventional typesetting for pages designed by graphic artists using manual tools. Bunnell was preparing to start his own desktop publishing magazine without us, and had even hired an editor.

Rather than be victims, we pointed out to Bunnell that his advertising manager had tried to get the subscriber list, and that fact was enough to warrant a charge of "willful misconduct" since it happened while Bunnell was preparing to acquire us. Our lawyer coached us to use the "willful misconduct" phrase. A day later, Patrick McGovern of IDG — the same media mogul who, together with his wife Lore Harp, co-founder of Vector Graphic, had helped us pick up our pennies from the sidewalk outside of the Computer Faire years earlier, and now Bunnell's corporate boss — sent a message through *InfoWorld* editor Jonathan Sacks that if the deal with Bunnell deteriorated, to give him a call and he'd arrange a deal.

McGovern's interest did the trick. We agreed to meet Bunnell halfway between our offices, which turned out to be the United Airlines lounge at SFO. "Besides parachuting bags of money into your backyard," Bunnell started out, clearly chastised by McGovern, "what is it that you want from this deal?" We wanted the money, but we also wanted a soapbox to argue the merits of desktop publishing technologies. We were granted the soapbox in the form of a column in the newly reformed magazine *Publish!* and the funds to publish a newsletter, the *Bove & Rhodes Inside Report*, which eventually became the *Inside Report on New Media*.

Reports from the inside

"Right this way," the stylish young woman from the PR department motioned us to a conference room in the bowels of Apple Computer. She pinned ID badges to our suit jackets. On the table was a cloaked device about the size of a paper trimmer from an office supplies shop. With a dramatic flair, the product manager unwrapped the device and introduced the Apple Scanner, which would be announced in three months.

As editors of the *Bove & Rhodes Inside Report*, we became analysts, thus gaining a level of status that enabled us to go to such non-disclosure-agreement (NDA) briefings "underneath the kimono" in the product labs of companies such as Apple, Adobe, and Microsoft. The press game for these companies was to convince analysts to talk to reporters in a certain way. The companies wanted the press to quote these analysts. Thus, analysts wielded great power. All of this amounted to *access*, an important advantage as computer-industry reporters relied on analyst quotes. As each new desktop publishing technology emerged, and even as each new product was introduced, reporters would call us for our opinion. We would keep track of our NDA end dates and respond (or not) to the press as appropriate.

We did not abuse this privilege by pushing one company's products over another's. We did not own stock in any of them, and did not work under consulting contracts or any other form of employment to them. All we asked of the companies that brought us into their briefings was to buy a $195/year subscription, the same as any reader. We thought that to be critical of an industry, one must not be financially vested in any part of the industry. We were wrong.

As it turned out, conflicts of interest were rampant on the analyst side of the industry. Newsletters and journals, and tech conferences, were chock full of such conflicts. One major venture capital firm publicly listed its "Family of Friends" which included some major newsletter pundits who presumably invested in the firm's clients. How can they be objective about companies that are either funded by the firm or that compete with companies that are funded by the firm?

Such conflicts were opportunities for analysts to cash in on the industry's momentum, and even steer the industry down a certain path. For example, the major pundit of publishing, Jonathan Seybold, also moonlighted as a consultant on the LaserWriter and PostScript for both Steve Jobs of Apple and John Warnock of Adobe. During the font wars of the late 1980s, he drew a consulting fee from Microsoft as well. And yet the *Seybold Report* claimed to be independent, objective, and critical.

And how far could a company get in the desktop publishing industry without paying to exhibit at the Seybold Conference? What does it mean when the leading industry analyst pits each company against the other, creating more standards wars and confusion for the consumer? It means more profit for the analyst who is also a conference promoter. There were no rules, only privileged players and loaded dice.

In other words we had nothing to gain and everything to lose by being critical. Let this be the lesson: With newsletters, it's not about the newsletter content (unless you are taking over from the late I.F. Stone). It's about the power a newsletter brings, in the form of advance briefings, exhibitions, and highly profitable and exclusive conferences.

Tony Bove moderated the Desktop Publishing Applications panel of Michael Tchong (ReadySetGo), Paul Brainerd (PageMaker), Ian Zerafa (FTL Systems), Royal Farros (T-Maker), and Ron Ford (Studio Software). February 15, 1986. Source: DGHealy (via Wikicommons)

In retrospect, it seems obvious that companies would have to pay the analysts in order to play the press game wisely. Back then, it seemed to be sacrilegious. Today, in the wake of executive-level scandals and high-finance politics, it just seems quaint.

Stuck inside of Word again

Although the companies touted desktop publishing tools as easy to use, learning to use them in real-life situations was like plunging into the heart of darkness. Trouble lurked in areas such as converting word processing files, downloading fonts, cropping and scaling graphics from other programs, sharing laser printers, managing style sheets and publication files, and preparing files for typesetting. There was also the occasional page that simply wouldn't come out of the typesetter.

Microsoft was a force to be reckoned with. Rumors swirled that Microsoft was working on a PageMaker killer that would also kill Ventura Publisher. The publishers of those packages were already hedging their bets by integrating with Microsoft Word. Microsoft was slowly taking over the markets for word processing and spreadsheet applications with products that were essentially clones of WordStar and WordPerfect (Word) and VisiCalc and Lotus 1-2-3 (Excel). However, the Microsoft killer page makeup application failed to materialize.

All during the late 1980s and early 1990s we relied on Word for writing and editing. We wanted to choke that paper-clip character that popped up in Word docs with inappropriate suggestions. Word's creepiness was directly related to the random suggestions it threw at you when you least expected them, and the way it corrected your grammar before you could finish the sentence.

A fanciful version of the annoying animated paper-clip help. (Thanks to Dr. Norman Clark in the Dept. of Communication at Appalachian State University, Boone, NC.)

Inappropriate help from Word

Full disclosure: After our newsletter gig ended and we were no longer writing commentary about the industry, we accepted a short consulting gig at Microsoft as documentation experts. We proposed a help extension that would play animations on top of the application screen to show you how to use it. Microsoft eventually turned down our proposal but adopted part of the idea with its animated paper-clip help.

Word was bloated beyond belief. You could customize Word menus so thoroughly that you could no longer find the spell-checker. All during our desktop publishing period we paid a tithe to the gods in Redmond for upgrades and even migrated to more powerful computers just to run new versions. When the software continued to behave erratically, upgrade after upgrade, we began to feel misused and abused.

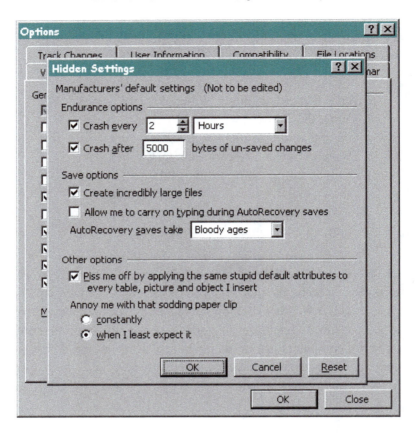

A fanciful version of the typical overly complex settings dialog. (Thanks to Dr. Norman Clark in the Dept. of Communication at Appalachian State University, Boone, NC.)

Using desktop publishing tools on Microsoft-based PCs was not pretty. Publishers preferred the Mac not only because it worked, but also because it was designed for graphics work, with a standard screen font format in the system, a simpler network that provided two-way communication with PostScript printers, and a standard SCSI port for exchanging portable hard disk drives. On the other hand, PC users were limited to screen font and width formats for their specific page makeup program. Popular

PC networks were not designed for two-way printer communication or downloaded fonts, and PC users didn't have portable hard disk drives to carry into service bureaus. These issues were the source of the rivalry amongst users.

The earliest explorers into this heart of darkness were the desktop publishing service bureaus that had to handle disasters unwittingly caused by their customers. In many cases the problem was a graphic element that couldn't be cropped or resized on the page, or a page that contained text in a bit-mapped font, which did not look good in print. Bit-mapped graphics and halftones processed by scanner software developed moirés when typeset unless they were scaled to certain percentages. When the service bureau couldn't typeset the page, they resorted to a form of alchemy to fix the problem, such as rearranging the page elements or shrinking some of them.

Word was essential for writing books, and at that time we were writing several books a year, including *The Art of Desktop Publishing* (Bantam), *Adobe Illustrator: The Official Handbook for Designers* (Bantam), and *Desktop Publishing with PageMaker* (Wiley). We used Word style definitions to define book chapters so that they could easily be sucked into PageMaker. Other programs, such as QuarkXPress, either ignored or mangled the Word style definitions, rendering them irrelevant and unreliable for our needs.

Word had planted itself in our world and threatened to outgrow anything else. Our clients sent Word documents that we could not open unless we also used Word. We sent Word documents back to them because that's what they wanted. The Word document format *itself* was the addiction. For a long time, the only way to open a Word doc file was to use Word; without alternatives, businesses migrated to Word and the rest of the Office package and got stuck there. As Word doc files proliferated (and as Microsoft wiped out the competition in word processing programs), they hooked everyone they touched.

Just what was in those files? More than you realize. At that time Word files could violate your privacy. The program was most often configured by default to automatically track and record changes you made to a document. A record of all changes was silently embedded in the doc file every time you saved it. It was easy as pie for someone to recover this record and see all the revisions. Most Word document files contained a revision log that was a listing of the last 10 edits of a document, showing the names of the people who worked with the document and the names of the files used for storing versions of the document.

A weapon of mass delusion

Word documents were notorious for containing tracked changes and revisions that could be embarrassing if discovered, and they were easy to discover. The British government of Tony Blair learned this lesson the hard way. In February 2003, 10 Downing Street published an important dossier on Iraq's security and intelligence organizations as a Word document— the same dossier cited by Secretary of State Colin Powell in his address to the United Nations. A quick examination discovered that the bulk of the 19-page document was directly copied without acknowledgment from an article in The Middle East Review of International Affairs, written by a student. As a result, during the week of June 23, 2003, the British Parliament held embarrassing hearings on the Blair dossier and other PR efforts by the UK Government leading up to the Iraq war. For the complete story, see "Intelligence? The British dossier on Iraq's security infrastructure" by Glen Rangwala.

Even today, you don't want to trust your most important documents to the Word doc file format. One reason is that files are extremely large. It seems like Microsoft stores the design specs and blueprints for the Titanic along with your text. These files are often many orders of magnitude larger than ordinary text files. Moreover, Word files are not secure — they can contain code that can destroy your computer. Word lets you create custom programs called macros for modifying your Word documents. Excel offers the same feature. But since Word and Excel let you save the macros along with the documents, and since it is possible to hide a virus in the macro code, these documents could easily be turned into Trojan horses carrying viruses. The code could be set to execute as the document opens.

How typical is this scenario: You receive a 20-page Word document from a client, and you need to print it. The document contains graphics — in particular, slides copied from PowerPoint right into the Word doc. But nothing comes out of the printer. Turn the printer off, restart, and wait for the system and printer to rethink everything... but still the pages don't come out. Hell freezes over, but the document won't print. Word docs are notoriously buggy when it comes to printing graphics, especially "objects" from other programs like PowerPoint, and even more so when those objects include text.

Somehow, something almost always went awry. Desktop publishing went from WYSIWYG to "You Can't Always Get What You Want" (with apologies to the Rolling Stones).

The reason for these and other printing and formatting anomalies was this: Word silently reformatted a document based on the computer's printer settings and fonts. This was bad news for certain kinds of documents, such as forms, that relied on elements precisely positioned on a page. In other words, Word documents were not guaranteed to look and print the same way on every computer and printer. The document's fonts may not have been available on another computer, and the substitute fonts forced the reformatting and caused pages to break in strange places. You can thank, among other things, the competing technologies for rendering fonts, which have befuddled the desktop publishing industry for two decades.

Tangled up in typeface

Printing problems can be traced back to the legendary Font War of the early 1990s, in which Microsoft and Apple faced off against the father of desktop publishing, Adobe. In this battle, Adobe represented the highly independent typeface industry by offering a high-quality format for typeface fonts, while Microsoft and Apple opted to use formats more convenient for their operating systems and HP laser printers.

A digital font is a mini-program that enables a system to display and print text with a typeface (such as Palatino) set to a particular size (such as Palatino 12). While Adobe didn't exactly invent this concept, the company *did* invent PostScript, and the Type 1 font format that works with PostScript. The combination enabled documents to be printed with high-quality fonts on different laser printers, and with even better quality using the same fonts on high-resolution imagesetters. This combination revolutionized high-quality printing in the late 1980s.

War broke out on the font frontier when Adobe published its Type 1 font specifications in 1990. This move democratized the world of digital type design, letting almost anyone make and sell their own digital fonts without needing a license or a massive computer system.

Adobe's font format dominated desktop publishing until Microsoft and Apple — strange bedfellows at that time — developed the TrueType format to challenge Adobe's dominance. Typographers weren't crazy about TrueType's quality, but even typographers had to eat.

In 1997, peace broke out in the form of a negotiated treaty and the OpenType initiative. OpenType allowed fonts to use either TrueType or Type 1 formats, and let a single font file work on both Mac and Windows, making documents much easier to interchange.

Back then and even today, Microsoft Word uses the fonts installed in your Windows or Mac system. When you first install Windows, only a limited number of fonts are available, but as you install other software, other fonts are added to Windows like new genes to the gene pool, and those fonts automatically become available to Word.

As we all merrily computed our way into the 21st century, our systems sprouted different fonts from all these different installations. When you created a document on one system, using its fonts, and then transferred that document to a different system, different fonts were substituted, with unpredictable results.

Conversion programs ensured that fonts in both formats would be ubiquitous. But when different fonts showed up in your system bearing the same name (for example, Palatino in either Type 1 or TrueType formats), your system and printer would get confused.

The font war confused even the typesetting experts. Maybe you could care less about fonts, as long as you got the document printed. But some of us wanted the printed document to look the same from one printer to the next. Thanks to the many differences in fonts and character spacing from one computer to the next, and from one printer driver to the next, you couldn't trust Word documents to look the same.

Even today, collaboration with Word documents can quickly get out of hand when others don't have the same fonts you have. Character sizes and spacing can change everything from the way lines break to the pagination and placement of footnotes. You might be referring to page 10 while someone else opening the document on a different system with different fonts would need to look at, say, page 9 or 11. And you can forget about trying to collaborate on complicated page layouts.

Giving PDFs a chance

The innovative PDF (portable document format) was originally made for collaboration. We covered the 1991 briefing when John Warnock, co-founder of Adobe Systems, first unveiled PDF to analysts. It was the final year of the Cold War and the Soviet Union had collapsed. Cooperation was in the air.

With a gleam in his eye and a silent nod to the analysts who had written extensively about the Font Wars that had nearly been his undoing, Warnock introduced PDF as if it were already a standard and read a statement that has since become the PDF manifesto:

"Imagine being able to send full text and graphics documents — this means newspapers, magazine articles, technical manuals, and so on — over electronic mail distribution networks. These documents could be viewed on any machine, and any selected document could be printed locally. This capability would truly change the way information is managed." (For an updated version of the complete story of PDF's introduction, see "How a tax form kludge gifted the world 25 joyous years of PDF" by Alistair Dabbs.)

This statement came true. However, we were promised the ability to transfer already laid-out documents from one page makeup program to another and *still be able to edit* the page elements. Page makeup vendors wouldn't come together to agree on a standard for page editing, only for page viewing.

PDF preserves the fonts, images, graphics, and layout of any source document, regardless of the application used to create it and regardless of what fonts you have in your system. PDFs include color profile information for more accurate color rendering across different systems. You can expect a PDF document that you create to look the same way on another computer as it does on your computer.

PDF solved many of the print problems plaguing the desktop publishing industry by encapsulating the font, graphics, and page layout information in a standard format everyone could use. By 1996, PDF had become a standard in the high-quality pre-press and color printing industry, and today, PDF files are ubiquitous as the *de facto* standard for distributing documents in a secure, reliable way.

Why not provide a format that people could edit? We asked Warnock this question in 1991, and he pointed out that the leading page layout programs, namely PageMaker and QuarkXPress, used proprietary formats, and these companies would never agree to cooperate at the editable document level.

This lack of editing turned out to be a feature, in the long run. Most people did not have the expensive tools for directly editing a PDF. You could encrypt PDF files for security, and distribute them knowing that the text couldn't be altered or copied without your knowledge. Today you can even digitally sign a PDF, making the format useful for electronic contracts.

With a track record of several decades, PDF has been adopted by governments and enterprises around the world to reduce reliance on paper. You already use it today to file your U.S. income taxes. It's the standard format for the electronic submission of drug approvals to the U.S. Food and Drug Administration (FDA) and for electronic case filing in U.S. federal courts. It is also the standard format used for advertisements in newspapers and magazines.

Sharing on a LAN

PDF made collaboration possible for desktop publishers, and desktop publishing as a cooperative effort became the best example for demonstrating how to link computers in a local area network (LAN) to share and exchange information, convert data from one format to another, and control access to files.

The first personal LAN to enter our home, in 1985, was the AppleTalk cable system that linked one or more Macs to an Apple LaserWriter. You could grow this LAN without too much trouble, adding more cables and Macs and using Centram TOPS or AppleShare to transfer files. With TOPS you could even designate one Mac with a modem as a communications hub and use the other Macs to connect to an online resource through the shared hub. Besides the bulletin boards described in the previous chapter, we would connect to CompuServe, The Source, Prodigy, and The WELL.

Before the LAN, the only way to share files was through what was called "sneaker-net" — running from desktop to desktop with floppy disks or hard drives. The advent of the LAN made it possible to sit still for a while and concentrate on work. Even so, America Online (AOL) seemed to go AWOL with more frequency as the deadline approached. The service had the annoying habit of holding up everything while it downloaded icons representing services.

In our 1987 book *The Well-Connected Macintosh* (HBJ) we breathlessly described how "you can, in fact, share files with people who are in different countries." It was important at that time to be able to replace old mainframe and minicomputer terminals with personal computers that could double as desktop productivity tools. The idea was to be able to extract large amounts of information from mainframe databases and publish them easily on a Mac.

Shining a light on the future

We were rockin' and rollin' in the publishing world, moderating "page makeup face-off" panels at trade shows, presenting tutorials on how to navigate through the heart of desktop publishing darkness, and writing our newsletter and a few books on the side, such as the first official book about Adobe Illustrator written with Fred Davis, a veteran tech editor.

It was a simple demo in early 1988 that turned a light on the future. We watched a computer-illiterate relative of ours learn how to use a Mac SE by playing with the interactive animated presentation that shipped with it (the Mac SE Tour, created for Apple by Animatrix of Palo Alto, founded by Marney Morris). Recognizing the power of animation and sound to capture attention, we became convinced that multimedia technology could spark new and different artistic efforts as well as more effective training material. Animation and sound could be far more effective than words in demonstrating how to use a program.

We had also been following the career of Gary Kildall, the pioneer who brought us CP/M. He was now working in a new medium called CD-ROM (compact disc — read-only memory), and in early 1985 his new company KnowledgeSet released the first encyclopedia on CD-ROM, *Grolier's Academic American Encyclopedia*. This was a step in the direction of Ted Nelson's hypertext, described in *Computer Lib*, in which readers could follow a link across a wide body of published works. Prototypes of interactive media, such as Apple's HyperCard released in 1987, did not go quite as far as hypertext, but let you move through predefined links in any direction through a text and graphics presentation, which was a major step in the hypertext direction.

We could see even then that CD-ROM as the storage medium would be enhanced by hypertext for publishing and accessing vast amounts of information. In our "Future Editions" column in *Publish!* in early 1988 we wrote about this topic under the title "Something is Happening Here, Mr. Jones", in which we speculated that the future of publishing would be hypertext:

> Hypertext publishing includes a promise of faster and less expensive methods for expressing new ideas,

transmitting them to others, and evaluating responses to them. Hypertext means customized documents in which text segments are threaded in a myriad of different ways, with browsing capabilities that can include filters to weed out excessive information. You can digress for quite a while in a hypertext system, following links into entirely new areas, and return quickly from whence you came.

Does the industry expect book and magazine readers to drop their pages and turn on their displays to read? Not immediately, but researchers will no doubt appreciate the ability to follow links through a vast network of data bases and writings... Although search parameters help us customize access to a database, we really want to browse through information quickly using an intuitive search pattern that can't be easily expressed in a logical and/or statement with keywords. We want to point at a reference and move to it (and to more references), then move back to the original text without losing our places.

The above prediction describes a web browser a full two years before the first web browser was invented by Tim Berners-Lee for the NeXT Computer in 1990, and *seven years* before the commercially available Netscape and Internet Explorer browsers. The very first web page would not be launched until 1991 by Tim Berners-Lee and Robert Cailliau.

However, the editors did not agree that CD-ROM and hypertext had anything to do with desktop publishing, which was about print.

For the subsequent issue we submitted the column again, but under a different title ("The Future of Text") that would not intimidate the editors. We explained that the skills required to publish electronically are a subset of those required for print publishing: writing and design skills, including the ability to put together pages. You don't need a printing press and color separation experts, but you still have to lay out information in a palatable form, using the screen as a page:

> Graphics and fonts do not disappear in this medium — on the contrary, the design skills of page makeup will be even more desirable for authors to have, since the software tools enable almost anyone to present information graphically. Graphics can save an enormous amount of time in aiding the digestion of complex information. Thus, desktop publishing and presentation tools are making better hypertext authors of those who use them.

After a second rejection, we enlisted publisher David Bunnell's help. This was, after all, our view of what the future held. We reiterated that the electronic medium was the next step, and that desktop publishing and presentation tools will make better hypertext authors of those who use them. Finally we pointed out the obvious, that all this content was already digital. Under Bunnell's influence, they finally published the column, which concluded:

> As more publications and presentations are made electronically, this vast body of published works grows automatically, and all we need to harvest it is the eventual standardization of text and graphics descriptions. The time is approaching when many of us will indeed read without paper.

The key lesson is that standards are necessary for establishing any new technology. With a standard in place, vendors and manufacturers can design tools and products that can work together, enhancing the experience for everyone. With text, graphics, and images, standardizing the formats turned into holy wars for dominating the design industry, but the standards that won these wars consolidated the strengths of the various technologies, making possible not only desktop publishing but also multimedia and internet publishing.

Text had already gone electronic with the ASCII standard, and as we mentioned earlier, we were already making progress with embedded codes for a meta-language for formatting text, similar to IBM's GML developed in the 1960s, which evolved into SGML of the late 1980s and the HTML we use today. Vector graphics remained mostly the province of proprietary formats used with the vector-drawing tools, but could be imported into PostScript.

Images, however, were a huge problem until digital image compression was developed. The size of a digital file storing a color image, with up to 24 bits of color information for each pixel (up to 16,777,216 color variations), was enormous for the machines of the 1980s. For many years we included color images in our publishing process by preparing photographic color separations on film, and integrating the separation films with the films for the printing plates. Color separation refers to the process of isolating the individual colors that compose an image so that it can be accurately reproduced on a printing press.

In 1986, Aldus of PageMaker fame introduced the Tagged Image File Format, commonly known by the abbreviations TIFF or TIF, to store digital images from scanners and cameras. The manufacturers began to take advantage of the format's complexity, implementing many of the tags to support varying subsets, which eventually led to the joke that TIFF stood for "Thousands of Incompatible File Formats".

To unite the various forms of digital image compression developed during this time, the International Standards Organization (ISO) started the process of creating the JPEG image standard in the late 1980s. "JPEG" stands for Joint Photographic Experts Group, the name of the committee that created the standard. JPEG specifies the coder/decoder algorithm, which defines how an image is compressed into a stream of bytes and decompressed back into an image. Since its introduction in 1992, JPEG has been the most widely used image compression standard in the world, and the most widely used digital image format.

No computer exemplified this new direction — supporting graphics standards — more than the NeXT computer by Steve Jobs, a stylistic and visually stunning black cube that ran a Unix-derivative operating system with the grace and poise of a Macintosh. By 1988 Jobs had suffered three catastrophes: the demise of the Apple III, the demise of the Apple Lisa, and the public relations disaster accompanying his firing from Apple. But he had learned how to court the computer press and not appear as brash and abrasive. And he led the industry by its ears into the mysterious twilight zone between personal computers and workstations — literally, as he demonstrated a duet with a symphony violinist that showed off the machine's CD-quality stereo sound output.

The NeXT computer: Steve Jobs again moves the entire industry forward.

Jobs declared that he had built the dream machine of the 1990s. The pinball wizard of the computer age garnered far more press coverage than his computer probably deserved, but the whirlwind introduction changed the computer industry forever. It turned the computer *itself* into a medium that displayed text and graphics well enough to read.

The NeXT cube also provided a path from print publishing to new forms of electronic media. For print publishers, the NeXT platform offered a dramatic improvement in graphics including a standard color management system and Display PostScript for the most accurate representation, at the largest and smallest display magnifications. For optical media enthusiasts, this machine included a magneto-optical drive that stored 256 megabytes, an order of magnitude higher than the 20-megabyte hard disk drives of the day. For future publishers, the cube included music and sound processing and built-in networking fast enough to run a networked version of Maze War, a 3D first-person shooter game.

My first encounter with the Steve Jobs Reality Distortion Field was in 1980, when he barged into my job interview with Martha Steffen at Apple and bluntly told me that Apple would own all copyrights of everything I wrote. I turned down Apple and went to Intel. My second encounter was after the Laser-Writer was introduced, when he explained, better than even Warnock could, the value of PostScript.

My third encounter was more substantive. During the NeXT period, Jobs was dating a close friend, Tina Redse, who had also worked with Jim Warren on our *Datacast* issues. We had dinner with Steve and Tina at one of Steve's favorite restaurants at that time, Flea Street Cafe in Menlo Park. (Despite it being his favorite, he still returned the pasta twice.) He agreed with us about how difficult the NeXT machine might be for a DOS or Mac user. When you dropped out of Windows, you ended up in DOS, which many people understood. You never dropped out of the Mac Finder, so you were treated comfortably. But if you dropped out of NextStep, you ended up in Unix, the most confusing system in the industry.

Our conversation veered off into one of Steve's favorite topics, the Beatles. We talked about different outtakes of songs available on bootleg LPs and CDs. We were surprised that Steve knew all about the "black album" of Beatles outtakes from the *Let It Be* sessions. The conversation ended with a prescient moment, in which we discussed with enthusiasm how we could use animation software to display an interactive version of the *Sgt. Pepper* album cover, with voices and words popping up as you moused over the members of the Lonely Hearts Club Band and listened to the music. We were dreaming about a new medium for interactive communication for text, graphics, pictures, and sounds, which became a reality a few years later (see the next chapter).

The NeXT cube was well ahead of its time, but costly. Most of its benefits relied on the existence of other NeXT machines in a network. Although it became a serious platform for color prepress service providers, small publishers who would have benefited the most from it could not afford it. Pursuing the perfect desktop publishing and training machine, Jobs had again moved the entire industry forward. It was no accident that Tim Berners-Lee used the NeXT computer to invent and host the World Wide Web.

Tools for the desktop press

The following are tools we used for publishing from the desktop.

Media: Books, magazines, newsletters, page advertisements

Tools and languages:

JustText, Aldus PageMaker, QuarkXPress, MacPublisher, Ventura Publisher

Microsoft Word, WordPerfect, Adobe Illustrator, Adobe Photoshop

CompuServe, The Source, Prodigy, The WELL, America Online (AOL)

Adobe Type 1 fonts, TrueType fonts, Bitstream fonts

PostScript, PDF creation and viewing tools

Linotronic PostScript imagesetter, Apple LaserWriter, HP LaserJet

Presenting Digital Media

Clarke's Third Law: Any sufficiently advanced technology is indistinguishable from magic. — Murphy's Computer Law

Nolan's Placebo: An ounce of image is worth a pound of performance. — Murphy's Computer Law

"Well, it's a crunchy peach pie / crunchy peach pie ..." Stuart Sharpe's stuttering animation of his girlfriend Jennifer introducing a peach pie raced across the screen and across our minds.

> *Well, it's a crunchy peach pie,*
> *And I went shopping for the peaches.*
> *Actually, I went shopping for all the ingredients,*
> *And found them all, except the peaches ..."*

Shots from "Crunchy Peach Pie", 1991.

We saw early versions of this animation live, on a Mac, in MacroMind CEO Marc Canter's living room in 1990. Marc Canter is a visionary who cofounded MacroMind in Chicago in 1984 with Jamie (then Jay) Fenton and Mark Stephen Pierce. The company subsequently developed MusicWorks, VideoWorks (written by Fenton, which evolved into the interactive animation tool Director), and MazeWars+, a multiplayer network game based on MazeWar. Canter was the first to christen the new technologies as "multimedia computing".

Sharpe's seminal animated art and music show appeared in *Gazette*, an Apple CD-ROM distributed years later at Apple Expositions and MacWorld Expos. You may recall, from the previous chapter, that we had seen the Mac SE Tour, created for Apple by Animatrix of Palo Alto — a curious mix of animated graphics useful for tutorials and training. Stuart Sharpe was pushing the limits of the same medium to create works of art.

As computers invaded the art world as production tools, they started to be recognized also as creative tools. Computer-generated graphics became a new medium for art, as well as a medium for business presentations. The term *multimedia* was reborn as a description of the combination of computer-generated animation, sound, and video.

Most importantly, this medium concerned the *computer screen itself*, as if the industry could see ahead more than a decade to the age of the iPhone and iPad. When The Voyager Company's fearless leader, Bob Stein, introduced photographic artist Pedro Meyer and his photo presentation at the Seybold Digital World Conference of 1991, everyone gasped. This was the first real example of dramatic, emotional multimedia. Meyer had added his voice and some music to a series of photographs and images played on an 8-bit color Macintosh screen. The subject matter was his parents, who were dying. He had taken pictures without developing the film until after his parents actually died, because he didn't know how to deal with the emotions. Then he put together this presentation.

No one could walk away from it. The deeply moving experience left the audience shell-shocked for over five minutes; you could hear a pin drop. In the awkward silence I thought first about my own parents (and felt guilty for not having called them in weeks), and then about how technology without emotion had no relevance in the typical day in the life of the human race.

Stein said at the time that it wasn't until he showed the Pedro Meyer presentation that Magnum (the largest photo agency) suddenly changed its attitude and realized that the future was the *display*. Pedro Meyer's computer show was the first major step into the future of virtual art galleries.

We can change the world

Graham Nash's "Chicago" was in my head as Cheryl and I explored the examples of new media in Canter's living room that day. The song from 1971 referred to the anti-Vietnam War protests that took place during the 1968 Democratic National Convention in Chicago, which we described in "Documenting in Print" — an event that indelibly marked me as a political idealist and an advocate of free speech. It occurred to us that if you had the opportunity to communicate an idea effectively, you could change the world, or at least your part of it.

"Visualization is really why computers were invented in the first place," wrote Canter in the foreword to our first book on Director, *Using MacroMind Director* (Que, 1990). "To help humans solve problems and communicate ideas, without having to fly to the moon or journey to the bottom of the oceans."

At the time we assumed that MacroMind Director would be most useful for storyboarding ideas for movie and TV scripts. We invited Lynda Weinman, then an artist working on ideas and concepts for the movie *Star Trek V*, to contribute examples to our book about Director.

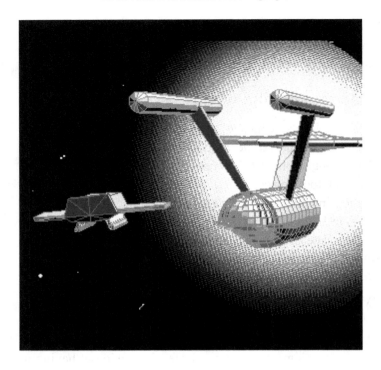

Where no Mac has gone before: Scene from animations created by Lynda Weinman for Star Trek V.

With multimedia, the printed page morphed into the electronic display. The term "real time" is extremely appropriate: displays made it possible to sustain communication over real time. The timeline in MacroMind Director's user interface was a direct representation of time. "Time is the insight that once you understand, you'll never forget," said Canter. "As represented by the [music-like] score in Director, time is the sort of idea that makes you say 'Oh yes, why didn't I think of that!'"

The Director score combined the direct manipulation of text and graphic "cast members" in space (on the screen) with a spreadsheet-like score to record movement over time. Professional animators marveled at the program's real-time responsiveness. The score would let you drag and select ranges of time as cells in a spreadsheet, and the stage area of the screen would let you place cast members as you would on a page.

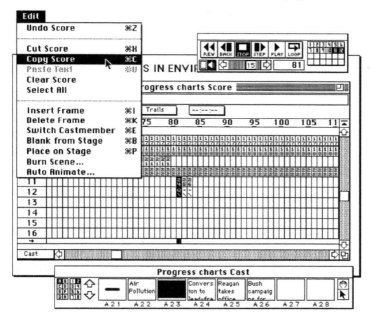

The score and cast in MacroMind Director.

Stuart Sharpe and others had contributed "clip animations" to the Director package that were short enough to use in low-budget projects. One of the projects we put together was a Mac floppy disk "greeting card" for Todd Rundgren's birthday, using (of course) his song "Birthday Carol" for the soundtrack.

Combining these moving images with music was a revelation. Apple had raised the stakes early with QuickTime in 1991, defining a digital video format that lasted nearly two decades. We were dreaming about a new medium for music combined with interactive liner notes, images, and videos. And we were fanatics of the Beatles. From *The New York Times* Technology column by John Markoff, Oct. 27, 1991, "Technology; Mouse! Movie! Sound! Action!":

> *For years, Tony Bove has been collecting bootleg music videos and songs by the Beatles and other groups. Now, by using programs based on a new Apple Computer software standard called Quicktime, which allows video and sound to be incorporated in Macintosh computers, he is developing his own presentations for friends to play back on their Macs.*
>
> *These creations are like music videos, but they are also "interactive," allowing the computer user to cut among songs, movie scenes, animated graphics and excerpts from documentaries in response to whatever is on the screen at any moment...*
>
> *"It's a grass-roots tool," said Mr. Bove, who edits a computer newsletter from his home in Gualala, Calif., north of San Francisco. "There has never been a technology that has been scaled down like this to the level of the individual."*

What prompted John Markoff to start his column this way was our private demonstration to him of our interactive version of the Beatles *Sgt. Pepper* album cover, which shows an assortment of celebrities — from Mae West and Bette Davis to Stan Laurel and Marlon Brando — as members of a "lonely hearts club" band.

The cover of the Beatles Sgt. Pepper *album.*

As we moused over each image while listening to the album's music, QuickTime movies the size of postage stamps would pop up to identify each member. The videos were small on purpose, so that we could play back the presentation from a CD-ROM disc.

We learned many of these techniques with MacroMind Director and QuickTime directly from a group of artists working in this new medium, including Stuart Sharpe, Joe Sparks, and Jim Collins, who took this medium to the next level.

Escaping Hollywood

If you walked the aisles of the digital media trade shows spawned in the early 1990s, mostly in San Francisco and Hollywood but occasionally also in Las Vegas, you might have been astonished to find Oscar-winning movie actors and Grammy-winning musicians leaning over your shoulder to gaze at the exhibits.

The entertainment industry had started to take notice of the superior quality of computer displays. At the same time, technology had lowered the barrier to entry, making it much easier for independents to make films, music, and games. The lure of Hollywood led some in the computer industry to be misled into thinking that the movie, television, and popular music industries would play a leading role in producing interactive entertainment and education for the next generation of multimedia player machines. But Hollywood in the early 1990s didn't have a clue. These industries were not as prepared as the computer game designers to express their ideas in an interactive form. Inexpensive computers gave people the power to interact with information in real time and follow tangents. At that time it was a new thing to be able to jump to a related topic by clicking on a link.

Hollywood was interested in using desktop tools for the production of real movies and real music, but the industry was still focused on these linear forms of entertainment. The successful interactive media titles would come from unlikely sectors — from new groups of artists and producers who were willing and eager to play with technology directly.

The first of these, Reactor, released *Spaceship Warlock* in 1991. It was a critical milestone in the progress of interactive media. A science-fiction epic designed as an adventure game, it offered fast animation and sound with a minimal storyline, casting you in the role of a wandering hero arriving at a planet dominated by the evil Kroll empire. By clicking the mouse you moved forward into this new world, acquiring credits so that you could buy drinks in the local bar, take taxis to spaceports, use the public videophone system, and so on. When confronted by characters, you could type a response.

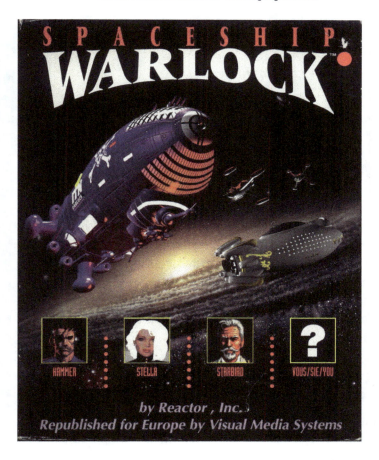

Spaceship Warlock *by Reactor (Mike Saenz and Joe Sparks).*

On our first excursion, we had to beat up a mugger to get enough credits to have a few drinks and call a taxi. The goal was to reach the end of the game and locate the hidden planet Terra. According to designers Mike Saenz and Joe Sparks, it took more than six hours to run through the entire interactive story even if you knew exactly what to do at each point. Some of the best animations were saved for the end, so that if you kept going you were rewarded for your efforts.

Archive.org to the rescue

You can download or play Spaceship Warlock online. The Internet Archive is a nonprofit devoted to providing universal access to quality information and offers several links to demo, Mac, and Windows versions.

A cool innovation for its time was the ability to save your progress in the game so that you could return at any time to where you left off. Spaceship Warlock was designed to be enjoyable, not to be as frustrating as some of the early adventure games, in which you had to remember to pick up an obscure object somewhere way back in the game in order to get through some obstacle.

Created with MacroMind Director, it required an Apple Macintosh II with a standard 13-inch (640 by 480 pixel) color display. The Reactor team used Paracomp Swivel 3D Professional for modeling and rendering, and Adobe PhotoShop for image retouching. All of the music was originally composed and performed by Sparks using the Opcode Vision sequencer with a variety of MIDI synthesizers, and a MacRecorder to digitize the music on a Mac.

The market for Macintosh-based CD-ROM titles was quite small at that time, but Reactor grabbed an impressive share. *Spaceship Warlock* was one of the best-selling Mac CD-ROM titles of all time, regardless of category.

Joe Sparks would go on to produce *Total Distortion*, another huge Mac best-seller. One of the most distinctive features of this music video adventure game was its sleep mode. When the player went to sleep, a dream appeared which consisted of a mini-game. If the player failed the dream sequence it would result in a nightmare.

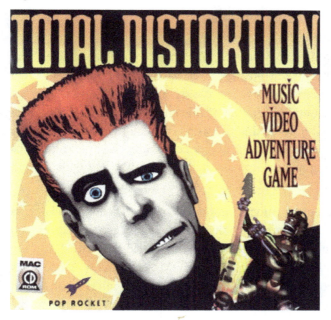

Total Distortion *by Joe Sparks (Pop Rocket).*

Roar of the dinosaurs

The CD-ROM industry had started with a big-bang conference in 1985. Seven years later it was still mired in confusion and suffering from poor sales and a small installed base. Bill Gates, chairman of Microsoft (which sponsored the first conference and the Multimedia and CD-ROM conference of 1992), acknowledged this situation in a keynote speech, stating that only about three percent of personal computers had CD-ROM drives connected to them.

At that time only the Mac was stable enough to support colorful multimedia titles on CD-ROM. With so many different Windows computers, the content publishers had to invest heavily to publish just one title, and had to develop custom tools to develop that title to work with different computers and game machines, further increasing costs.

Most of the titles seemed like wild shots in the dark. Their publishers took it to heart that, like a much smaller version of the movie industry, their business was based on hits that kept them afloat after many misses. The clutter of useless CD-ROM products was a natural byproduct of this entrepreneurial activity. Perhaps it is not so ironic that the worst of the lot were published by new media divisions of the larger companies, or by partnerships with those companies, to cash in on this new wave with content they already controlled, such as movies and games.

One exception came from the desktop publishing industry, as you learned in the previous chapter. *Verbum* magazine developed a $49.95 "multimedia magazine" on CD-ROM entitled *Verbum Interactive*. Billed as the first CD-ROM periodical, the disc emulated the look and feel of a magazine, with an interface in which you used your mouse to turn electronic "pages" to update the computer screen.

The first disc (a 2-disc set) offered numerous examples of multimedia art and animation, with rock songs by Todd Rundgren, Graham Nash, D'Cuckoo, and other artists, which could be played on ordinary audio CD players, or played on CD-ROM players so that you could click on interactive animation to play the music, and see animation in sync with the music. The other disc featured a roundtable discussion of multimedia by a panel of experts, with sound and simulated (six frames per second) video of each panel member answering questions. You could select a question and see responses from each panel member, as if you were accessing that person's mind directly, by clicking the mouse.

Verbum Interactive was demonstrated in 1991 at multimedia conferences as an example of what software-only QuickTime video decompression would look like (digital video in a small window). It may have been the first example of a low-priced multimedia content product supported in the traditional magazine way: with paid advertising. In this case, the ads were animated demos from companies such as Adobe Systems, MacroMind, Paracomp, and Letraset.

It may be hard to imagine, in this age of digital commerce and on-demand media, that the most innovative entertainment products of the 1990s were units of merchandise encased in plastic packaging that had to be sold in physical markets or through the mail. Due to the limited shelf space in retail stores, large publishers would bid for more shelf space and squeeze out the smaller presses, which were relegated to mail-order sales. The publishing businesses consolidated brands by reprinting backlist catalogs and throwing new media titles into glossy catalog-style CD-ROMs to see what would stick.

"We are risk averse," one investor told us. "We only want to invest in something that is 90 percent finished and is a sure thing, and then, we want to retain control." However, there was no evidence that the financial community knew anything about creating the next hot product. In fact they all relied on garage band-inhabiting, debt-laden wild-eyed software fanatics.

Given their tendency not to invest in new, creative titles, the threshold for entering this business was low for publishers, but high for entrepreneurs like ourselves. It costs about $3-4 per shrink-wrapped disc to manufacture and package the disc in small quantities (including a plastic jewel case and a page of liner notes), to sell a CD-ROM like a music CD. Larger companies' budgets enabled them to encase their CD-ROMs in a large colorful outer box or other expensive package, to make their product visibly stand out and dominate the limited retail store shelf space. The one-time authoring and production costs for a title were in the range of $50,000 to $150,000, not counting rights licensing fees (if any). And marketing was far more expensive because it was difficult to precisely target and reach the right audience.

Microsoft, the elephant in the room, was supposed to come to the rescue. When the company announced the Multimedia PC in 1991, prominent Mac-based developers jumped on the bandwagon to get their titles ready for the Windows-based computers. Heady optimism was in the air at the grand opening, staged at the American Museum of Natural History in New York City in, appropriately enough, the Hall of Meteorites. Microsoft's message was "Bring the world to your senses," but critics of the company's forceful licensing practices joked that it should have been "bring the world to its knees."

Microsoft's meteorite, the MPC, ended up cratering the multimedia industry by introducing too many choices. The confusion undermined its potential to provide a unified vision for multimedia computing. Low-priced slow CD-ROM drives flooded the market as consumers were lured in with bundled encyclopedias. Vendors were free to add non-standard audio cards that were skimpy on features. Titles would not function properly on certain CD-ROM drives or with certain audio cards. After spending all that money for a supercharged PC with an audio card and a CD-ROM drive, spending even more for a title that didn't work was a slap in the face. Creators of impressive multimedia titles like *Spaceship Warlock* could not get their titles to work in that nonstandard operating environment, so they stayed with the Mac.

Compounding this confusion were risky investments in other CD-based formats, such as Philips' CD-i and Commodore's CD-TV. Philips subsidiary AIM dictated strict terms for investing in content — too strict for most of the talented professionals at that time. As for CD-TV, the early units were not selling because the salespeople didn't know what they were or what they could do.

Learning to interact

CD-i showed some promise with music titles that you could "interact" with, in terms of choosing what visuals to see while you listened. Of these, the most interesting was Todd Rundgren's 14th studio album, *No World Order*, released on CD-i, making it the first interactive album in history. The interactive version (still available at Archive.org) includes the ability to alter the playback of the music by selecting a predetermined sequence by either Rundgren or one of his four guest producers.

"Up until *No World Order* I had been pretty anal about what the audience was supposed to experience from my records," wrote Rundgren in his autobiography *The Individualist*. "It was a revelation to confront the fact that they were looking for a more granular way to absorb media, one that conformed to the new age of portable devices."

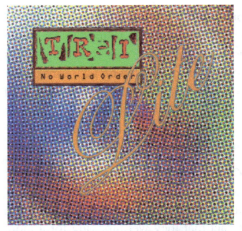

Todd Rundgren's interactive CD-i, No World Order.

Rundgren and his software partner Dave Levine devised algorithms to enable a listener to navigate a library of sound clips in real time and use primitive artificial intelligence to select clips automatically. But these inventions were short-lived, as Rundgren explained: "Shortly after we delivered the final product Philips began to phase out the format so we ported it to Apples and PCs, a historically tortuous process especially on the Microsoft side since every PC was a random collection of junk hardware and crap system software. Suddenly we were in a nightmare world of customer support." That took all the fun out of making these types of products, so he moved on.

The Mac was easier to develop for, because at that time it was nowhere near as confusing as the mass-market PC. Apple had also developed the QuickTime architecture to provide a software-only digital video format that could play videos from CD-ROM discs. For these reasons, the Mac CD-ROM platform was stable enough to support colorful multimedia content titles, which is why Reactor's *Spaceship Warlock* rocketed up to the top of the charts. Despite the small size of the installed base, innovation in multimedia technology and progress with the acceptance of multimedia technology were occurring faster on the Mac side of the market.

It's amazing that Bill Gates, the world's richest man at that time — worth billions — could not figure out how to make the MPC a success, even though the Mac was an example to copy. It took the mavericks of much smaller companies, mostly under financial duress, to come up with solutions. In the shadow of large companies like Apple and Microsoft, the much smaller MacroMind pioneered the concept of an independent multimedia player that could run the same CD-ROM content on either a Mac or an MPC. At one memorable moment, Marc Canter and Stuart Sharpe shared a keynote stage with Bill Gates to show off MacroMind's multimedia player. "We've got two screens up on the wall. One is a Mac – it's got these animations flying around the screen. The other one is this black screen with the command line. Bill Gates goes, 'I'd like to announce multimedia coming to the PC on the new Windows platform.' He hits the spacebar and the animation flies across the Mac screen over to the PC screen. It was an awesome moment."

But nobody's making money

As the industry struggled with the definition of multimedia, anguished over the opportunities for a vibrant multimedia market, and deliberated over the many legal roadblocks that prevented the use of established content in new, original works, not many were making any money. When IBM's Peter Blakeney moaned about the Mac-centric audience at one of these Hollywood conferences, L.A. Times columnist Michael Schrage interrupted with "Hey, no one wants to hear IBM whine." Blakeney shot back, "I may be whining but my pockets are jingling." And yet, at that time the only reason IBM's developers could make money was because IBM was paying them or subsidizing their efforts with free equipment.

As we worked the corridors of trade shows looking for funding for these "extremely risky" CD-ROM projects, we came away feeling slimed by those who were hoping to profit as soon as possible. At this embryonic stage, we thought the revolution should be nurtured, not tortured with improbable sales projections and endless meetings with investors who did not understand why product development timelines were so long and expected unrealistically swift returns on their investments.

"We had kept ourselves alive by taking our product and building in-store marketing demos and simulations," explained Marc Canter. "That was a work-for-hire business and we were able to bootstrap. We were producing things; that's how we made the product [MacroMind Director] better. We'd have our programmers in the booth of MacWorld. The customers would come up, they'd ask them a question, and the programmers would go back to the hotel room, add a feature, and come back the next day with the feature." The key piece of the technology was a multimedia player for interactive animation. You didn't use the authoring environment to distribute the product; you could author it once and play it back on multiple platforms.

At a Stewart Alsop Agenda conference, with Timothy Leary in the background hobnobbing with the real Ben and Jerry of ice-cream fame, Canter and his crew in the Ritz Carlton hotel room shared joints and Jolt Cola into the night to develop the Stewart Headroom animation, with Alsop's head on the cartoon Max Headroom's body, then applied the same technique to create a John Sculley Headroom. "I would hit the keys on the keyboard," said Canter, "and John Sculley's mouth would open and close, and we would put words in there."

This kind of manic production brought out the innovative quality and efficacy of the tools. Service businesses were making a living, and nearly 85% of the CD-ROM titles were created with Director, but Canter knew they were leaving money on the table. "There's real confusion. Nobody really understood multimedia; they were looking for the money. You know where the money was? Games. So the game industry benefited from it."

Fantastic Voyager

One publisher stood out as different from the rest. The Voyager Company's founder, Bob Stein, defined the gold standard of electronic books on CD-ROM by combining text and graphics with the music that readers could interact with, and in the process created a new, much better way of writing about music.

Imagine how difficult it would be to learn to play the guitar from only text and how much easier it is to learn from a hands-on, one-on-one session with a teacher. A multimedia presentation can condense a complex subject to make it more palatable and allow endless replays to help the reader get over the first hump of comprehension.

Our voyage into multimedia CD-ROM titles began with animated presentations derived from some of our instructional books. The idea was to prepare an electronic version of a book combined with animated demonstrations and information. Thus, we contributed to the Expanded Books Project run by Voyager during 1992, which investigated how a book could be presented on a computer screen in a way that would be both familiar and useful to regular book readers.

Our role, once again, was to explain how to use these tools, and our first expanded book was about using MacroMind Director to create animated presentations. Some of the highlights of that book were animated versions of storyboards in our printed book from Lynda Weinman, who had worked on the *Star Trek: Enterprise* TV series. This electronic book also helped introduce the clip animations by Stuart Sharpe.

Stein had pushed the idea of an encyclopedia on the Atari platform before engaging in a partnership called the Criterion Collection to produce classic movies on laserdiscs. The Criterion *Citizen Kane* and *King Kong*, released in 1984, had a direct influence on the future of media consumption by using multiple audio tracks to provide commentary to support the films. Criterion also included deleted scenes and mini-documentaries on the making of the films, which are now considered standard movie industry practices.

Apple's introduction of HyperCard in 1987, which offered the card metaphor for combining text, images, and sounds (and eventually video clips), changed everything for Stein. Within weeks of its release, engineers had figured out a way to use it to control laserdisc players. While laserdiscs had been connected to computers for custom training applications, Stein and company put HyperCard in the mix and created the *National Gallery of Art Laserguide*, a tool for navigating and exploring the National Gallery collection.

While this was quite innovative, the laserdisc was still an analog medium, and using it with a computer meant dealing with two screens — one connected to the computer, the other to the laserdisc player itself. The obvious solution became CD-ROM, and Voyager started with music. In 1989 the company started working on *Beethoven's Ninth* CD-ROM for the Mac. This and a handful of other products were the first to define the genre of "multimedia titles", using CD-ROM as a distribution medium.

Beethoven's Ninth served the industry as a reference point for CD-ROM educational titles, and as a result, music publishers and performers started paying attention to CD-ROM as a music medium that could also present electronic liner notes. An interactive, illustrated exploration of Ludwig's *Ninth Symphony*, the title provided the music and the historical context in which it was created, and explanations of musical concepts and instruments used in the piece.

Beethoven's Ninth was written by Robert Winter, a professor of music at UCLA and a leading authority on Beethoven. The entire *Ninth Symphony* was recorded on the disc, and you could read Winter's commentary as you listened. "I had no experience with programming before starting," Robert Winter explained to us in 1992. "It was an experience somewhere between writing *War and Peace* and cartoon captions. On the one hand, everything has to relate to everything else — you can't have sloppy logic. But it is like cartoon captions because a well-designed screen 'page' can't have much text anyway — you can get only one idea on a page. It was the most challenging writing I have ever undertaken. It was also the most rewarding."

From this title we learned that Beethoven's "Joy Theme", the most recognizable piece of music in the *Symphony*, was composed to the words of Friedrich Schiller's "Ode to Joy" and was also used in Stanley Kubrick's film "A Clockwork Orange". We also learned a great deal about the use of percussion instruments in the *Symphony*, and the program played musical passages to illustrate musical concepts. You could stop and start the music and read the text on the page that described the section you were hearing. You could read the vocal portions of any movement in German or English, see the musical score for various parts, and play the MIDI version or the CD version. The multimedia version of *Beethoven's Ninth* depended on accurate timing to synchronize the display with the music, so much of it was copied from the CD-ROM disc to your computer to make it work properly.

Voyager's titles competed with Warner New Media's *Funny* and *The View From Earth*, Brøderbund's *Whole Earth Catalog* and *Living Books: Just Grandma and Me*, ICOM Simulations' *Sherlock Holmes*, PIMA's *Escape from CyberCity* and *Caesar's World of Gambling*, and Compton NewMedia's *Encyclopedia*. Voyager soon added titles for children including the innovative *AmandaStories* and the wonderfully ironic *A Silly Noisy House*.

Interact with the Voyager title sampler

Interact with a huge download at Archive of The Voyager Company's 1992 diverse title sampler developed for Windows and ported to the Windows 3.11 emulator.

A silly noisy workplace

Voyager's workplace was a communal space in an old building right on the beach (on the beach side of Hwy. 1) in Santa Monica. In 1991, Voyager hosted the last party of the Seybold Digital World Beverly Hills Conference, overflowing the two-floor headquarters of The Voyager Company in Santa Monica (just north of the pier on the Pacific Coast Highway).

Before the party I was given a tour of Voyager's operations, and I wondered how this eclectic group of video experts, programmers, and multimedia authors could work so intensely, in such a wide-open space of cubicles, and be so close to the beach. This company sat on the edge of the wild and imaginative multimedia content titles business, its headquarters sticking out of Santa Monica like a little toe testing the rough Pacific. These beautiful people liked what they were doing, which was escaping from reality by creating multimedia content titles. Which, of course, made the beach seem an afterthought, a useful sunny space for exercising.

Dominating the large kitchen (for what is a commune without a large kitchen?) was a large hairy man named Hugh who was, at first, bellowing about the merits of Apple's QuickTime technology, but had somehow veered off into a lengthy description of the monster scene he made whenever he attended his local Coastal Commission meeting to blunt the ever-expanding wedge of L.A. growth in the northwestern hills. I imagined this wild-haired barefoot programmer barging into a meeting of local conservatives, bad-rapping the land developers, yet able to leap effortlessly from one discussion on Cajun cuisine to another on ham radio data transmissions, and then to still another on the strengths and weaknesses of HyperCard versus other tools, while pausing to comment on someone else's description of the trails in Yosemite.

Voyager's fearless leader was Bob Stein, soft-spoken but never at a loss for words as a speaker at multimedia conferences. Standing on a balcony overlooking the beach, next to a home-brew propane barbecue that belched flames and looked like it would explode at any minute, Bob Stein exhibited that night the same casual but polite deference when talking to us, or to a particularly nosy public relations expert, or to a notable computer industry newsletter editor, or to his young daughter, or to Christopher Cerf when asking him if he'd like to edit a multimedia presentation about the Persian Gulf. His self-effacing manner was contagious — after talking with him we felt important and yet at ease. He was one of a handful of people in the industry who understood the effective communication power of interactive content.

At that time our first son was only two, but he loved *A Silly Noisy House* because it was so interactive. Children could discover clever animations, nursery rhymes, riddles, songs, and many different sound effects by pointing and clicking on furniture, appliances, bathroom faucets, doorways, and other familiar objects. Author Peggy Weil created an environment that encouraged kids to play and explore at their own pace.

Bob Stein's vision stayed within the confines of what was practical for the marketplace. According to Peggy Weil's website, "Bob [Stein] invited me to create whatever I wanted with the following conditions: it would have to be a product for children, it would have to be predominantly audio, and it would have to require no licensed properties. These requirements were strictly business; the emphasis on audio rather than visual reflected that high-quality audio could be produced at lower cost than quality animation. He specified a children's product because it was both evergreen and a broader demographic than any specific topic for adults. There was simply no money for licensing." The title made *The New York Times* All Star List in 1992 and won the Silver Medal in the 1993 *New Media* INVISION Awards.

A Silly Noisy House by Peggy Weil was one of the very first interactive titles for children. The Voyager Company released it on CD-ROM in 1991 for color Macs.

AmandaStories, the very first point-and-click program for kids, appeared on the Mac as black and white animated stories on floppy disks, then as CD-ROM titles. *AmandaStories* won many prizes, including Macworld magazine's SuperStacks Award, a Parent's Choice Award, and PCM Magazine's Publisher's Pick Award. The stories were popular around the world because they communicated through sound effects and animation, not with words. They were warm, funny, and original, and each story contained many branches, offering children different alternatives and new opportunities to exercise their imaginations every time they played. Six stories were about Inigo, a spunky kitten always getting into mischief, and the other four stories were about Your Faithful Camel, a traveling animal with a magic suitcase full of tricks. On each screen page of each story there was at least one invisible button that performed some action, such as displaying animation, playing sounds, or moving to other pages. A button might create a different effect depending on which buttons were clicked previously.

The Voyager Company would go on to publish the best CD-ROM ever, *The Residents: Freak Show*, designed by Jim Ludtke. Designed for the artist collective known as The Residents, this title elevated multimedia into high art, influencing future CD-ROM music-related titles such as Marc Canter's MediaBand and Ty Roberts' *Jump* for David Bowie. The latter, produced by Roberts' company ION in 1994, lets you edit your own version of David Bowie's "Jump They Say" video and mix your own version of the music.

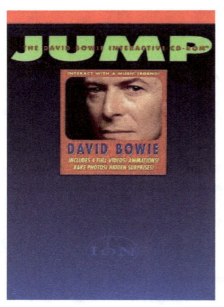

David Bowie's Jump *interactive CD.*

Diversions

Those who can dance are considered insane by those who can't hear the music. — George Carlin

Voyager's Digital World party was still ramping up when I decided to leave for yet another party. Someone had slipped me DMT, and I was flying for about an hour while standing motionless on the curb of the Pacific Highway in front of the building.

It gave me time to think about the 1960s counterculture and its relationship to the new media industry. John Markoff's excellent book, *What the Dormouse Said: How the Sixties Counterculture Shaped the Personal Computer Industry*, documented how the Sixties experience with psychedelics played an early role in the development of high tech, including Englebart's mouse. He described the evolution as all of a piece: the drugs, the antiauthoritarianism, and the messianic belief that computing power should be spread throughout the land. "It is not a coincidence," Markoff wrote, "that, during the Sixties and early Seventies, at the height of the protest against the war in Vietnam, the civil rights movement and widespread experimentation with psychedelic drugs, personal computing emerged from a handful of government- and corporate-funded laboratories, as well as from the work of a small group of hobbyists who were desperate to get their hands on computers they could personally control and decide to what uses they should be put."

You can see the influences of the psychedelic experience of strobe lights and plasma light shows in the electronic mandalas of the emerging virtual experiences. Our notions of community involvement stem from the grassroots organizing of the Sixties, and led to the experiments mentioned earlier, including the Community Memory Project in Berkeley, the People's Computer Company in Menlo Park, the first computer games, and Ted Nelson's vision of hypertext which ultimately evolved into the World Wide Web.

You can chart the intellectual progress from the early Merry Prankster experiments with sound and multimedia, Stewart Brand's *Whole Earth Catalog* (promising "access to tools" with a photo of the entire planet), and Charles Reich's interview with Jerry Garcia, *A Signpost to New Space,* through Brand's excellent *Cybernetic Frontiers* and Ted Nelson's *Computer Lib,* to Dr. Timothy Leary's writings in the Eighties, and the emergence of thousands of desktop-published 'zines and magazines covering this new frontier such as *Verbum, Mondo 2000, Processed World,* and *Wired.*

Would there have even been an Apple, a Microsoft, or a Google, if not for the cultural influences of the Summer of Love? "What would America be like," mused Allen Cohen, the editor of the *San Francisco Oracle,* "if we had somehow gone from the gray flannel Eisenhower-McCarthyite Fifties to the three piece suit Reagan Eighties without the interceding influence of the beat, hippie, civil rights, women's rights, farmworker's rights, and anti-war movements? The result of such a time warp probably would have been a direct line, without much resistance, to fascism or even holocaust."

Steve Jobs credited the psychedelic experience in making him more enlightened, according to Walter Isaacson's biography *Steve Jobs.* "I came of age at a magical time," said Jobs, quoted by Isaacson. "Our consciousness was raised by Zen, and also by LSD.... Taking LSD was a profound experience, one of the most important things in my life. LSD shows you that there's another side to the coin, and you can't remember it when it wears off, but you know it. It reinforced my sense of what was important — creating great things instead of making money, putting things back into the stream of history and of human consciousness as much as I could."

Eventually one of my friends from the Digital World conference picked me up and we went to Timothy Leary's place. Lots of folks from the conference showed up at the good doctor's home in the hills above Beverly Hills for a party that started at 11 p.m. A note on the unlocked door said "Don't knock, just come in."

I had known many of these partygoers for years. I had bumped into some of them an equal amount of time at computer trade shows and at Grateful Dead shows. There was a considerable overlap, although it didn't make sense on the surface of it: here were people who excelled in an exact science, the science of programming, where bits are either ones or zeros, no ambiguity, but you could find them twirling in the tie-dye swirling craziness of a Dead show, swishing and jiggling to the primitive African beat mixed with unpredictable electronic feedback, breathing the pungent aroma of pot smoke, dressed in loose scarves and shorts, no underwear, no bras, bopping to the improvisational music. Music with no right or wrong notes or mistakes — just lots of random ambiguities and mushy emotional outbursts of peace and love.

They all hugged each other in a kind of hippie greeting that caused stares from the suits back in the conference hall who were thinking, look at those hackers, reveling in anarchy, nearly practicing group sex on the trade show floor, they're the crown jewels of the industry, they get founders' stock, they get invited to the right parties, they get the backstage passes! While we struggle and toil and wear ties and close deals just to make a decent living... Was it some kind of yin-yang thing? Exactitude balanced with disorderliness? It's no joke, or perhaps it's a cosmic joke, that the NASA space program's software was written by Deadheads hooked on "Dark Star" and "Mountains of the Moon".

Leary's house was not out of the ordinary for the neighborhood, with its sweeping view of the L.A. basin, nice furniture, and an awesome audio and video rack, but it did have strange equipment attended by nerds lurking in back rooms, a lot of cushions in the living room, and photos of beautiful actresses and models posing with the good doctor in his prime. Leary himself was in one of the back rooms involved in an experiment, laying on a gurney stoned out of his gourd while hooked up to biofeedback equipment. I would eventually have a frank talk with him in the early morning hours about using some of his archival material for a CD-ROM project about the San Francisco counterculture in the 1960s.

From left: Tony Bove, Ty Roberts, Marc Canter, Ann Greenberg, and John Eric Greenberg.

The next day Leary showed up at the conference, and I took him back to our Hilton suite to show him the progress we'd made with our project. He headed straight for the honor bar and, ever the Irishman, drank over a hundred dollars worth of booze. We were sharing the suite with the pioneers of interactive media. Marc Canter, still CEO of what had now become Macromedia, was pitching the *MediaBand* CD-ROM title, which combined the many talents working in Director with a group of modern musicians working in electronic music. Also sharing the suite was Ty Roberts of ION, pitching a David Bowie title, *Jump*. Many of

Hollywood's innovative producers and directors were taking a peek at our demos.

Our next stop was Las Vegas, home to Liberace, Wayne Newton, and Howard Hughes' ghost. That year's SIGGRAPH conference could not have been hotter in 110-degree Vegas. SIGGRAPH was the only show in which the Unix nerds of the military-industrial complex mingled with the Unix nerds of the movie industry. Guess what they had in common! Unfriendly interfaces, mathematically-prescribed 3D models, otherworldly surface maps, and $30,000 software packages running on $20,000 workstations linked to million-dollar Crays.

Vegas was the perfect place for this type of conference. Inside the auditorium were flybys and surface renderings of the planet Venus; outside it was Frankie Avalon singing "Venus". Inside it was George Coates Performance Works presenting Invisible Site, a live multimedia performance with projected interactive real-time computer imaging; outside the fake volcano out in front of the Mirage Hotel erupted every fifteen minutes, stopping traffic on the Strip. Inside we were treated to glimpses of virtual reality including virtual dining (EAT by Michael Naimark & Company), Silver Suzy on a surfboard (Chris Walker of MR FILM), and video games in which you could throw real objects into virtual space (David Thiel, Incredible Technologies); outside we had virtual unreality on the Strip with $1.95 all-you-can-eat breakfasts at the casinos, giant water slides at Wet World, and the Magic Motion Machine at the huge Excalibur Hotel.

The Virtual Reality room provided comic relief from the aisles of serious mips and mflops, serving as a sort of daycare center for the technologically overwhelmed adults and the children of the space age. We stayed there most of the time beating the virtual drums of Vincent John Vincent's Mandala VR System, throwing real cue balls into the Match Five video atrium, flying through protein molecules. I held my own in the NASA Ames Virtual Wind Tunnel, and rode the virtual tramway in Karlsruhe, Germany. In the Electronic Theatre we watched two hours of animation while holding wood sticks with plastic reflectors (they looked like giant tongue depressors) to illuminate tiny pixels on a vast screen representing the entire audience (set up by Loren Carpenter of Pixar). At the George Coates performance we wore plastic-and-cardboard 3D glasses as if it were the 1950s and watched an estranged individual try to delete his name and address from the vast data banks of the universe. I asked George if he continually encountered glitches with the multimedia machines (as we've come to expect from demos and press conferences), and he replied that the glitches are part of the performance — the random elements help to make the audience think the performance is real.

With the exception of the midnight tearoom hosted by Coco Conn in one tiny corner of the mind-boggling Caesar's Palace, the usual nightclub parties were boring — it seems that all of the gin-and-tonic salespeople were at the nightclubs looking to score with booth bunnies. To drink with other journalists, we went to the Computer Graphics World party at the Liberace mansion and wandered through the fabled bed and bathroom of the late great piano player, only to find a Liberace look-alike poised to take our picture with a poodle. How can we miss him when he won't go away?

The full moon cast an unearthly glow over Death Valley as we escaped from Nevada and plunged ever onward in search of the perfect multimedia show. But the next one that impressed us was neither virtual nor digital. Post-beatnik poet and *San Francisco Oracle* editor Allen Cohen had put together and presented live at the Fort Mason theater, a 600-image slideshow, personally curated and richly narrated, about the Haight-Ashbury district in San Francisco and its role in the Summer of Love in 1967. We would soon produce this slideshow and narration as a multimedia presentation on CD-ROM.

Haight-Ashbury rise and fall

"It was a marriage made in hacker heaven — the compact disc and the personal computer." Thus wrote Charles Bermant in his Technopop column "CDs Put a New Spin on Your PC" in *Rolling Stone*, Issue 638, Sept. 3, 1992 (p. 77):

> *A more ambitious project still far from completion is* San Francisco in the Sixties: The Rise and Fall of the Haight-Ashbury — *eight hours of music, film, graphics, and poetry. Says producer Tony Bove: "This allows you to examine a body of work in many different ways and to explore the material from your own perspective." Bove, a computer-industry columnist and record collector, first toyed with multimedia technology while experimenting with various Beatles records, bootlegs, and film clips, creating a point-and-click mystery tour. Now he is pitching the idea to the powers that Beat. He feels a project like a multimedia Sgt. Pepper has breathtaking possibilities, from identifying each personality on the cover to using rehearsal tapes to trace the development of the album.*

The Beatles management never got back to us. We also tried to contact Yoko Ono about doing a title based on "The Ballad of John and Yoko". But Allen Cohen's presentation, accompanied by a facsimile edition of the *Oracle*, resonated with authenticity, and Cohen's narration was a compelling historical perspective on the hippies and the Haight-Ashbury experiment. As students of rock music history and folklore and avid readers of all the books on the subject, we were impressed by Cohen's depth of analysis and interpretation. As the organizer of the original Human Be-In in 1967, he approached the subject matter with a certain poetic grace and critical eye.

Today, young people flock to San Francisco from all over the planet. They do it to participate in creating a new world, to demonstrate their inventiveness, to learn from other like-minded entrepreneurs, to be a part of what is hip and stylish, to get rich quick, and to rub shoulders with and breathe the same air as their heroes in high tech.

Nearly sixty years ago, young people made a similar pilgrimage to San Francisco to participate in creating a new world. They demonstrated their inventiveness at building community. They learned from other like-minded entrepreneurs of the spirit. They *became* hip and stylish. They worked together with their heroes in rock music, pop art, the new journalism, and street theater.

The Summer of Love in San Francisco in 1967, centered in the Haight-Ashbury, was far more influential than most people realize. The generations that followed it have taken for granted the social and political changes wrought during that pinnacle of the Sixties. The civil rights movement, the free speech movement, the ecology movement, anti-war, anti-nuke, and peace movements, the psychedelic movement, black power, women's liberation, and gay liberation — all forged in the crucible of the Sixties — were the foundations for the social and political environment we live in today. As talk of revolution continues to spread after each of the recent presidential elections, we are reminded of the last great revolution in this country, when the Sixties peace movement, fused with the summer of love generation, stopped the Vietnam war.

Today, we refer to a fun gathering as a *love fest*. We talk about being in a *groove*, or *uptight*, or *burned out*. We *tune in* to the media, and cool gadgets *turn us on*. We *drop out* of the rat race from time to time, but we can still *pass the acid test* to demonstrate something that is working. *What a trip!* All this terminology has its roots in the Summer of Love. Our ultra-casual dress in our high-tech offices — everything from headbands and blue jeans to tie-dye shirts, caftans, sandals, and self-styled non-conforming garb — echo the style and culture of the hippies of that period.

Cohen recalled that the movement was like "a children's crusade that would save America and the world from the ravages of war.... Peace and love weren't just slogans but states of mind and experiences that we were living and bearing witness to. Living in harmony with the earth was an ideal that we felt and perceived as real experience. We were bringing forth a second Renaissance that would change human culture."

The Summer of Love was a pivotal moment in the emergence of true compassion for other human beings, inspiring future generations to work on improving all of our lives. "The beat and hippie movements brought the values and experiences of an anarchistic, artistic subculture, and a secret and ancient tradition of transcendental and esoteric knowledge and experience, into the mainstream of cultural awareness," explained Cohen. "It stimulated breakthroughs in every field from computer science to psychology, and gave us back a sense of being the *originators* of our lives and social forms, instead of hapless robot receptors of a dull and determined conformity."

Turn on, tune in, and drop... out!

Timothy Leary's famously short speech at the first Human Be-In in 1967 kicked me awake. In early 1992 I had watched many hours of film and video footage at the Manchester, UK studios of Grenada Television, looking for clips to use in a project we were starting with Allen Cohen. It turned out that the Brits had done the best documentaries about the Sixties, even though many of the Sixties events took place in America. Not only did the British documentarians capture the highlights of the Summer of Love in the Haight-Ashbury, but they had also interviewed rock stars like Paul McCartney and Paul Kantner about their experiences during that time. These, and many other clips, were available to use in "optical media" for an exorbitant flat rate.

Dr. Timothy Leary at the Human Be-In, 1967.

We had been entranced by a 1987 documentary by John Sheppard for Granada Television, narrated by former Beatles publicist Derek Taylor, titled "It Was Twenty Years Ago Today", which told the story of the Beatles' *Sgt. Pepper* album, which made its debut in 1967. We learned from the director that the documentary had been a labor of love and an extremely complicated exercise in procuring the rights to the material. The clips in the film seem to come from everywhere, but he pointed us in the direction of the BBC and Granada. At that time the BBC had no idea how to charge for the use of its content in "optical media" as it was called then (in later years the organization would license all of its content for use on the Internet). So our sole source of useful video content turned out to be Granada Television, along with a short clip from a CBS documentary about the Haight-Ashbury.

We wanted to bring together these experiences: the live slideshow experience with documentary clips and still images, and the contemplative experience of studying the artwork, images, and texts of the *San Francisco Oracle* pages. We believed at the time that all musical and film-related entertainment would one day be distributed with art, lyrics, interviews, and simulations for customers to interact with, and this belief has mostly come true.

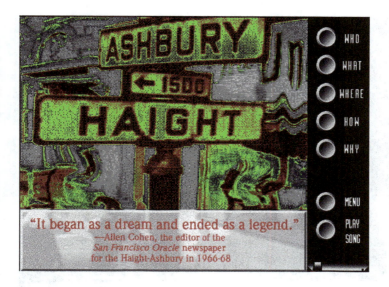

Early 1993 demo of the Haight-Ashbury in the Sixties *CD-ROM, highlighting the media's potential for supplying art, lyrics, interviews, and simulations with music albums.*

The final self-paced presentation included artwork, articles, still images, poetry, and commentary from the original *San Francisco Oracle* newspaper for the Haight-Ashbury, which was published from 1966 to 1968. You would be immersed in the title simply by navigating through its content, using a colorful animated timeline to jump to different periods, or assembling a mini-presentation by dragging topics. We brought together authentic voices: Allen Cohen with his gritty New Yorker streetwise beat-poet voice, and Raechel Donahue, the queen of FM rock radio KMPX (and later KSAN) in San Francisco before the suits took over. And we included authentic music from the Grateful Dead, Tom Constanten, Jefferson Airplane, and Big Brother featuring Janis Joplin.

Following page: The poster for Haight-Ashbury in the Sixties *by Rockument.com in 1996, with original art by Alton Kelley.*

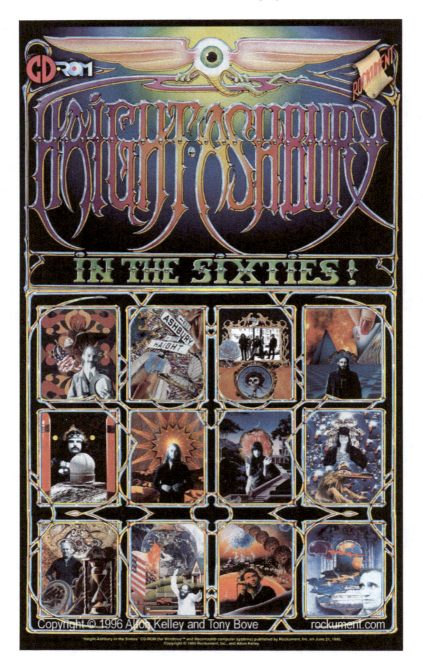

Roach papers

Kelley had us collect and provide him with our joint roaches, so that he could unravel them and use the spent roach papers in the art. Look carefully, they're everywhere!

With over 100 photographers and artists in Cohen's slideshow contributing over 700 images, getting the rights to use this material took more than three years. We bucked the trend toward signing up all rights and left out all the intimidating language suggested by lawyers. We offered prorated royalties and $30 per image as an advance for the *non-exclusive* right to use the images.

With video clips from Granada and CBS identified and priced, we had to move more quickly to license the music before anyone else could change their minds and their prices. Groveling at the fringes of the hugely lucrative entertainment world, we learned right away that deals were confidential. You couldn't trust the numbers you read about in *Billboard* or *The Hollywood Reporter*. There were "standard" deals that nobody signed, and there were real deals that were kept secret. Mostly there were no music deals for "optical media" or anything resembling digital media because no labels or publishers had figured out at that time how to price their content.

We did the uptown turnaround back-step too-de-loo at many offices in Hollywood searching for the right person to do business. One label after another turned us down or didn't even return our calls. Then we hit upon the right strategy: We focused our negotiations on the Grateful Dead as the anchor for the title. Years of obtaining backstage passes at shows and playing in the Graceful Duck (see Chapter 5) helped cement our relationship with the Dead's business side, and thanks to super-connected Deadhead friends like David Jacobs, a VP at MacroMind, we were able to get the rest of the band's attention. Even Jerry Garcia took a look at our prototype.

The package design by psychedelic poster artist and "visual alchemist" Alton Kelley sealed the deal. Kelley was the co-founder of the Family Dog, and Mouse Studios partner of Stanley Mouse, who had participated in creating some of the Dead's album cover art and logo.

For 45 minutes worth of music, the Dead charged us the same royalty rate that they would normally make off an album. The material they licensed to us for our CD-ROM would eventually be released as a 2-CD set titled *Two From the Vault*. They would get essentially the same royalty from both products. However, it was significant that the Dead established a rate of about $2 a unit for 45 minutes worth of music; as a result, RCA agreed to license Jefferson Airplane's "White Rabbit" based on its duration of about two and a half minutes, Tom Constanten agreed to license his music at the same rate, and Dave Getz, drummer for Big Brother, sold us an early take of Janis Joplin doing "I Know You Rider" at the same rate.

We bought the non-exclusive right to include the performances as background music accompanying narration, and sequenced with images and animation. This was a critical sequencing right, similar to licensing for a movie or TV show, because music performers and publishers are concerned about what appears on the screen while their music is playing.

Flush with success from the music world, we shopped the Haight-Ashbury title around to several CD-ROM publishers before deciding to publish it ourselves, as an affiliate of a distributor, rather than using an existing publisher. One reason was the absurdly low royalties offered by CD-ROM publishers, from which we'd have to pay out to license the music and other content. Then we found out how truly dumb some of these publishers were with regard to the music business. One CD-ROM publisher and distributor, a leader in the game industry, asked if it could issue the audio tracks of Grateful Dead music to the record-buying public, even though such a product would compete directly with Grateful Dead Records. For budding multimedia publishers at the dawn of this industry, their lack of perspective was startling.

Using these existing publishers was unattractive not only due to the lack of sufficient royalties, but also because they left us with an uneasy feeling about their interest in maintaining a relationship with the content owners. What if they used a rock group's image in advertising in a way that was not appropriate? The lack of control over the use of the content owner's material and image was an important issue for this title.

So for these reasons, we would not turn the project over to a publisher. We already knew about another title in which subtle, complex interactive animations were without warning stepped on by huge, muddy, corporate feet. The publisher authorized the removal of important scripts without informing the programmer. Pieces of the CD-ROM were created by different teams, and stitching these pieces together became a nightmare for everyone involved. Disc performance suffered. And so on. The problem was a lack of a true project director, who functions like the director of an independently-produced movie: faster than an idling corporation, more powerful than an indecisive bureaucrat, and able to spend tall budgets with a single bound. I was determined to play that role for the Haight-Ashbury title.

The animated title screen for Haight-Ashbury in the Sixties *by Rockument.com. Images in the title flashed by quickly in the center circle, accompanied by the opening drums of the Grateful Dead's "The Other One".*

We eventually struck an affiliate deal with Compton's New-Media of Carlsbad, California, "a leading producer of interactive information, infotainment, edutainment, and entertainment software on CD-ROM." Compton's had made its name with a CD-ROM encyclopedia, and had gone on to sign over 20 affiliates to distribute over 150 titles to more than 7,500 retailers throughout the U.S.A.

As an affiliate, we supposedly had total control over the contents of the discs and Compton's had no claim to any of it. However, the company forced us to put a splash screen on it (a credit screen that appears as the title loads), which was incorrectly set to the wrong color palette for Windows PCs. While the splash screen had no effect on the title running on a Mac, it prevented the title from running properly on a Windows PC. The cost for us to resolve this support issue for irate customers was astronomical. However, it was also satisfying because we showed customers how to change their display settings to properly experience the story as it was designed to be viewed.

Compton's NewMedia advanced enough money to produce 50,000 units and send out royalty advances to the 300+ content licensees. To accommodate the music, videos, two-hour documentary sequence, and game, we produced a two-disc set. To create a distinctive package that could double as a joint rolling tray, we bought 50,000 blank cigar boxes and printed Kelley's artwork on them. We also included a poster detailing the 12 digital "cards" in the game.

"Digitize everything!"

Allen Cohen was skeptical. All of the content for the *Haight-Ashbury in the Sixties* project was in what we call "analog" form. He looked at all the old photo prints, slides, posters, videos on tape, his story on audio cassette, and so on. Pages of rainbow-colored ink from the *S.F. Oracle*. Psychedelic art on wall-sized canvases representing the 12 "cards" of the game. Even the package and poster art by Alton Kelley started as large canvases with collages of hand-placed and hand-painted prints. A poet from the 1960s, Allen didn't understand how we could accomplish this presentation without compromising quality.

"Simple," I shouted with glee. "Digitize everything!"

This was not a fantastic leap of faith on our part. We already knew about Kodak's Photo CD format, launched in 1991 and adopted by entrepreneurial photo-finishing services as an option for developing film. Photo CD was a key enabler at this point, as it could store about 70 high-resolution images on a single disc. Nearly 1,000 pieces of artwork would be scanned in some way, even the large canvases, using cameras on tripods, and then developed directly onto 12 Photo CD discs. We would then copy the digital files from these discs to use in the animated user interface.

The Apple Macintosh was designed from the beginning to play digital sound, and various sound recording and editing products were available to create a digital version of Allen's narration.

Music was a different problem. We needed technology that could play both CD-quality music (essentially 16-bit digital audio) and animation from the CD-ROM, or settle for less-than-CD quality with 8-bit digital files. While 8-bit was fine for the narration, it would not do for music.

The answer came from Ty Roberts and his company ION, which was working on CD-ROMs for artists including David Bowie and Todd Rundgren. His plug-in for Macromedia Director enabled us to interleave 16-bit (CD quality) audio with animation on a CD-ROM disc so that it could play back from the disc.

We also used the Cinepak encoder and decoder (a.k.a. video codec) with a Macintosh II-based VideoVision card from Radius to capture the analog video from SMPTE time-coded half-inch Betacam SP videocassettes. We digitized the video into the recently developed MPEG digital video format. We had to keep the size of the digital video picture small in order to play it back directly from the CD-ROM disc.

You can't test capability without actually burning a CD disc and playing it. At that time, CD-ROM "burners" (recorders) were around $2,500, and blank discs were about $10 each, so it was no big deal to burn discs for testing. Armed with these quickly-burned CD-R discs, we set out across America on a press tour.

Meet the press

Timothy Leary helped us introduce the *Haight-Ashbury in the Sixties* CD-ROM at Michael Gosney's Digital Be-In #7 on January 6, 1995. The press tour and conference road trip included television appearances on San Francisco morning news shows and interviews with tech journalists for most of the major computer-related magazines, including *Wired*, *PC World*, *MacWorld*, *PC Computing*, *The Wall Street Journal*, and *The New York Times*.

The official T-shirt for the Haight-Ashbury in the Sixties *press tour and road trip in 1995, by Rockument.com.*

I admit that all the attention (especially of my hometown paper, the *Philadelphia Bulletin*) went to my head, and I violated the first rule of press interviews to promote products: Don't ever disagree with the interviewer.

But first, here's a perspective on the press attitude:

- *WIRED* called it "An unflinching, non-judgmental chronicle."
- Stephen Manes of *The New York Times* wrote on July 18, 1995 that it "illuminates the era with the fuzziness of a lava lamp."
- According to Scott Rosenberg the next day in the *San Francisco Examiner*, "Other histories are more comprehensive, but it's hard to imagine a more pungent introduction to the cultural flow of '60s San Francisco — from Beat-era stirrings to the high hippie days, then through disillusionment toward the New Age retreat of the '70s. Each strand of the counterculture is represented, from utopian psychedelia to Eastern mysticism, from back-to-the-land communalism to off-the-pigs militancy... When 'Haight-Ashbury in the Sixties' inspires such then-and-now connections, it's fulfilling any historical work's highest calling."
- Even Playboy magazine weighed in: "Now the users can do the Sixties without any of the bad acid."

So it was surprising to have a famous tech journalist react as if the title was radioactive. Walt Mossberg, then of the *Wall Street Journal,* greeted me at the paper's offices in Washington D.C. in a work shirt, jeans, and sneakers, with slightly balding semi-long hair, looking every bit like my leftwing older brother who'd been a veteran of the civil rights movement.

Mossberg immediately pointed to a green leaf in the decorations around one of the opening screens, and pronounced the CD-ROM unfit for human consumption, or at least for him to review in the *WSJ*. "This promotes the use of drugs. I wouldn't want my teenage daughter to see it."

"But... but... Forget the leaf, you'd have to know already what that is!" I stuttered a reply that name-dropped many of the intellectuals of that period who had provided content in the title, from Allen Ginsberg and Alan Watts to William S. Burroughs, Ken Kesey, Timothy Leary, and Gary Snyder. "Are you saying you wouldn't allow your daughter to read them? Or to hear 'White Rabbit'?"

By this time Compton's PR rep was pointing at his wrist-watch and interrupting about making the next appointment in an effort to drag me out of the office. I continued to waste time arguing with Mossberg, who thought I shouldn't publish the title until the industry accepted a warning sticker for CD-ROM titles along the lines of the Parents Music Resource Center (PMRC). "Did you have this same attitude about the underground press in the Sixties?" I yelled back. "The ones the IRS bankrupted under orders from the FBI?" And so on. Never mind that Mossberg could have done this over the phone and saved us the trip to D.C.

As luck would have it, the next interview took place in my room at the Waldorf in New York City, with two guys from UPI. They were quite enthusiastic about the CD-ROM. One of them pulled out a bent, crinkled joint, and asked sheepishly if it was okay for them to smoke it during the demo.

Mossberg was clueless about the need in the CD-ROM industry for something a bit more entertaining than an encyclopedia, and the need for small-time entrepreneurs to get a leg up with their first products. The title injured his middle-of-the-road sensibilities. Besides, who really knew what his daughter was up to when he wasn't looking! Nevertheless, we should go bankrupt, because we didn't paste a warning sign over the Kelley artwork. I wonder what he would have thought about Michael Gosney's review, which talked about "Dead licks, incredible graphics, history, games — and comes packaged in a cigar box for convenient stash management."

The money-go-round

The money goes 'round and around and around
And it comes out here.

— The Kinks, "The Moneygoround" (Ray Davies)

Today, in the age of Amazon, YouTube, and Apple Music, it's hard to imagine the marketing cycle for a physical content product such as a CD-ROM. In the first place, the market itself was just a minuscule part of the larger computer market. CD-ROM titles were expected to justify computer system upgrades, and had to essentially pull their customers into the new age, forcing them to buy CD-ROM drives, audio cards, and larger color displays.

You had to generate press attention well in advance of release, so that retailers would be primed to order more units, which then had to be manufactured and shipped to the stores. The advertisements and press outreach didn't occur close enough to the point of purchase. Pricing had to be established so that you could stay in business while letting retailers keep 60 percent; most CD-ROM titles in the retail channel were $19.95. Retailers could return units they didn't sell, to open up shelf space for other titles. It could take over three months to get your first dollar from sales. Affiliates such as ourselves, in the form of Rockument, Inc., had less clout with retailers than the bigger publishers, and were likely to be last on the list to get paid.

And so it came to pass that we went broke, thanks to our affiliate relationship with Compton's New Media. First, the company spent $40,000 of "our money" on a wasteful advertisement in *Rolling Stone* magazine that mostly promoted them and came out way too early for sales. Then Compton's was acquired by the Chicago Tribune Company, which promptly invalidated affiliate deals. These moves left us without any payment for sales and a product stuck in the retail chain.

As we discovered when we went to the Milia festival in Cannes, even European distributors wouldn't touch a product already sucked into the maw of U.S. retail. All we could do in that town was to commiserate over cognac in the lobby of the Le Grand Hotel with John Perry Barlow, lyricist for the Grateful Dead and co-founder of the Electronic Frontier Foundation (EFF), and Douglas Adams, author of *Hitchhiker's Guide to the Galaxy*. Adams remembered our *Inside Report* and in particular the issue in which article titles (such as "Living in the Digital World") were puns of George Harrison songs. "I know George," he said. "He'd find it funny."

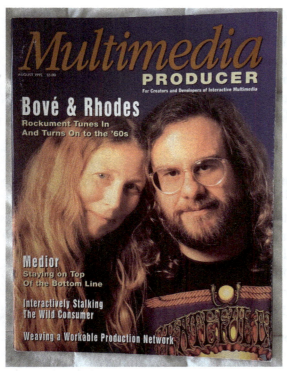

We were the subjects of a feature article on multimedia production for the first issue of Multimedia Producer *magazine in August 1995.*

We weren't laughing, however, when we learned that there would be no compensation from Compton's for selling over 40,000 units, at $19.95 apiece. Not a cent. This was not a good way to start a CD-ROM rock music history company.

Rather than getting to do new titles such as Swinging London, or the Ballad of John and Yoko, we ended up doing the Ballad of Tony and Cheryl. We dissolved the company and got separate jobs, but we did retain one media property. Rockument.com, created in 1995 as one of the first internet websites, is still alive today and streams videos from the Haight-Ashbury CD-ROM.

Everlasting value

If you examine the relationship of art with commerce, you find that it is not a struggle between good and evil, but a struggle to build something of everlasting value. Art and commerce are codependent; neither exists without the other. Art exists only if people are provided the opportunity to experience and appreciate it, and only through commerce can people see, hear, smell, feel, or taste it.

We spoke with hundreds of customers who called us for well over a year after Compton's shut down, answering calls at all hours from upset people seeking tech support to properly set their screen resolution and restore the color palettes that Compton's had screwed up. They were mostly grateful and very satisfied with the CD-ROM experience. Customers shared with us how much they truly enjoyed our labor of love, even if they had picked it up for $5 at a flea market.

You know you've offered something of value when you get feedback. The following are just three out of hundreds of messages gathered from customers browsing our website that validated the Haight-Ashbury CD-ROM as an important work of history:

I needed this. My wife says I live in the past. I always say, so what. I don't think my values have changed since the 60s but it's hard to live in the 90s with 60s ideals. Thank you.

I'm looking for a picture of Janis Joplin. Can you help me? I work with some youngsters who have never even heard of her, much less seen her.

I am so glad that I have stumbled onto this web site. I just handed in my final paper for my Social Movements of the 1960s class. One of the questions had to do with Haight-Ashbury and the counterculture as a political movement. I look forward to finding more information about the 1960s on the web!

That last comment is telling: "I look forward to finding more information..." The web gives you the power to plunge further into the depths of knowledge and information. CD-ROM titles could never go far enough. Works of art, works of the heart, and other sorts of niche CD-ROM titles simply didn't make enough money to justify the effort on economic terms. The few titles that broke through this barrier to get attention and sales were the rare exceptions that we embraced with enthusiasm — starved as we were for content that was truly interesting and relevant to our lives.

Anticipating the future jump into cyberspace, we wrote in our newsletter in 1991 that multimedia production was getting better all the time. "If we find new ways to distribute multimedia information — whether it travels down a wire, bounces off a satellite, or is merely treated as entertainment or software rather than as a book in the distribution chain — then perhaps anyone can make a living creating interesting content for publishing. Freedom of media? Let it be."

Along this winding road from desktop publishing to multi-media we learned that the issues of quality, control, and artistic freedom are essentially the same. The tools we used for adjusting images for video, for example, were the same ones we had used before, for preparing images for printing. Multimedia added another dimension to instructional content at least a decade *before* YouTube.

Revolutions, whether in art, business, or matters of state, create a new world only by sacrificing the old. It is no different with media inventions. The production techniques borrowed from desktop publishing, such as page design and color imaging, fueled growth in jobs and opportunities as designers, typographers, and photographers moved from print to the web. As a result, the decade of the 1990s saw nothing less than the eruption of digital media at the expense of paper, of websites at the expense of newspapers and magazines, of online shopping at the expense of bookstores and music emporiums.

Tools for presenting digital media

The following are tools we used for publishing animated and musical projects in digital form.

Media: CD-ROM titles, digital photography, digital video

Tools and languages:

Adobe Photoshop, Macromedia Director, Swivel Pro, SoundEdit

PowerPoint, Persuasion, Photo CD, CD recorders, slide recorders

HyperCard, Lingo

Publishing Online

Instant karma's gonna get you
Gonna knock you right on the head
— John Lennon, "Instant Karma"

Murphy's Law of Thermodynamics: Things get worse under pressure. — Murphy's Computer Law

In the 1990s we entered a cannibalistic phase of the new media revolution, in which the media industry began to eat its own tail in a frenzied ouroboros contortion. Publishers and media conglomerates unleashed a torrent of indiscriminate investments into digital media that all but obliterated their existing print media of newspapers and magazines.

By 1994 we had already sacrificed our careers as respected industry analysts to join the frenzy of producing multimedia projects. We were dragging physical media projects, such as carousels of slides, wall-sized posters, and analog videotapes, across the great divide into the indestructible digital world, using every skill we had learned while doing desktop publishing. And while we were gathering our roach papers for Alton Kelley's Haight-Ashbury CD-ROM package design, and dreaming of producing a series of "rockumentaries" that could be distributed online, a great industry upheaval occurred.

Jim Clark, a successful Silicon Valley entrepreneur who had started Silicon Graphics (SGI), and his recruit Marc Andreessen, a software engineer and developer of an early web browser, had become paper multimillionaires in a single day, as a result of their extremely successful initial public offering (IPO) for Netscape, in which the stock's value soared from $28 to $75. The company introduced the Mosaic Netscape browser for the World

Wide Web, which eventually became Netscape Navigator. The foundation, a way to navigate and browse the web, was now in place to completely transform all media technologies.

Around the same time, Stanford graduates Jerry Yang and David Filo established Yahoo (an acronym for Yet Another Hierarchically Organized Oracle, although the founders would insist that they just liked it as the slang term for a rude, unsophisticated, and uncouth Southerner, or as the race of fictional beings in *Gulliver's Travels*). They curated what eventually grew into a gigantic set of bookmarks for the web. Yahoo grew rapidly through the 1990s until eventually the company lost its competition with the Google search engine.

Also around the same time, Jeff Bezos started an online marketplace for books called Amazon. It was probably not even the first online bookstore, but it soon became the largest and is now a behemoth that sells just about everything through an online interface. When Amazon filed for its IPO in 1997, Bezos reported that daily site visits had grown from 2,200 in December 1995 to 80,000 in March 1997, and revenues were already above $15 million and growing at nearly 3,000% per year. By 1998, Amazon had 3.1 million customers and hundreds of millions of dollars in revenue.

We really could have put the web to good use as a distribution channel when we launched the Haight-Ashbury CD-ROM, but it was too early for Amazon and e-commerce. Book publishers initially saw the web as a way to peddle their paper and CD-ROM products. Eventually they would find themselves in competition with their own distributors, who were using the web to fulfill bookstore orders.

During the 1990s shelf space in stores was increasing at a much slower rate than the rate of development, which meant that publishers were begging for shelf space, and interactive CD-ROM titles, especially niche-oriented ones, sat in warehouses unsold. Ironically, the consumers were looking for niche-oriented titles, and they complained that the titles for sale were similar to each other and didn't offer enough depth.

The web was slowly coming to the rescue. Publishers could offer a sample (albeit a sluggish experience) of a CD-ROM, which stores couldn't offer to everyone who walked in, not even using an interactive kiosk. Another was the unlimited amount of shelf space, which acted as a sort of leveler — all products were equal on the web. However, the business model was different. Publishers were still essentially merchandisers, with books or CD-ROMs as the merchandise, in a world where the lowest price ruled. The model favored the larger publishers who could routinely throw money into advertising.

To break the merchandising model, publishers needed to find a way to get content into the people's hands simply and elegantly, and somehow monetize the process. The way to do that was to commit not to paper, tape, or CD, but entirely to digital.

It's the page, not the paper

In 1988 we predicted that content publishing and distribution would move online, utilizing a cross-platform page makeup format that would evolve from commercial applications. In the years that followed, all efforts to create such a format failed. Then along came the World Wide Web with HTML (HyperText Markup Language), which redefined the "page" as a computer screen display medium.

HTML evolved from structured document standards that (are still) widely used for generating PDF files. Tim Berners-Lee, inventor of the web, first described HTML in late 1991 as a set of elements, or tags, strongly influenced by SGMLguid, a documentation format in use at CERN based on the Standard Generalized Markup Language (SGML). DocBook SGML and LinuxDoc are examples of documentation created with SGML tools. As technical writers, we were already versed in SGML, which evolved from Generalized Markup Language developed by IBM in the 1960s.

Central to the concept of a web page is the ability to *link* to other web pages, so that contextual elements are available around the stuff you are reading if the author chooses to provide them. As a reader you can, for example, follow links to definitions, histories, background data, and feedback, as well as footnotes.

Ted Nelson's vision for hypertext was the model for browsing pages online. Hypertext supports very complex and dynamic systems of linking and cross-referencing. The most famous implementation of hypertext is the web itself, written in the final months of 1990 and released on the internet in 1991.

Many first saw the web as a giant encyclopedia. Professionals in many industries, including doctors and lawyers, had already learned to search for information online rather than in books. Now they could follow links to more information, turning the web into a gigantic pool of source material for footnotes and digitally stored repositories for research for research topics. In our previous roles as technical writers, we saw the web as the perfect way to provide ongoing, updatable documentation that was as easy to browse as paper but without printing and distribution costs.

However, for linking, HTML is a one-way ticket. Ted Nelson had envisioned hypertext as a two-way system, in which you could link forward *and* backward to create and follow associative trails, and had even devised a way to measure royalties to provide revenue streams for content creators. While the web proved to be flexible for adding one-way links, it was not designed to create two-way paths. It was, however, possible to use buttons and images as links, which enabled the design model of HyperCard on the web, and to provide buttons for purchasing and feedback.

As a result of the web's sudden popularity, HTML became the "typesetting" language for online publishing. We perceived it as a populist revolt over the hegemony of Microsoft Word, PageMaker, Ventura Publisher, Adobe Illustrator, and other publishing industry giants. HTML evolved as a standard in the hands of the people who were rejecting the established norms. It was a revolution of structure over form, and with HTML, document structure could be preserved without having to worry about formatting. No longer could Microsoft lock you into using Word for your documents.

Microsoft eventually had to support HTML as an export option for Word. Even Bill Gates at first hadn't comprehended that the "killer app" for the internet had arrived, and by the time he did, it was out of Microsoft's control.

"Just Say No to Microsoft"

In the mid-1990s, Microsoft was trying to unite the industry behind Windows 95 as the operating system for the future. But an onslaught of malicious hackers and the planting of viruses sidetracked this effort, and the critics pounced. "The idea that Bill Gates has appeared like a knight in shining armor to lead all customers out of a mire of technological chaos neatly ignores the fact that it was he who, by peddling second-rate technology, led them into it in the first place," wrote Douglas Adams, author of *The Hitchhiker's Guide to the Galaxy*, in Douglas Adams on Microsoft (*The Guardian*, Sept. 1, 1995).

Microsoft would issue security updates that were like putting fingers into the holes of a leaking dike. The updates didn't reach enough people fast enough to stem the flow of malware. The Mafia couldn't have thought up a better protection racket than the platform Microsoft provided at that time: an architecture loaded with loopholes that criminals could use to gather information, exploited by "virus protection" companies for profit.

Well aware of Microsoft's history of locking customers in with new features at the expense of security, proposing pseudo-open standards, and spreading FUD (fear, uncertainty, and doubt) about its competitors, I decided to write Just Say No to Microsoft (No Starch Press) to set the record straight.

I also wanted to point people to alternatives to Microsoft, such as Linux for an operating system, and OpenOffice to replace Microsoft Office and Word, especially for creating documentation.

An extreme reaction by a Microsoft employee to Tony's book Just Say No to Microsoft *(No Starch Press).*

When I first downloaded the free OpenOffice.org package, I felt a strange new sense of elation. There were no restrictions. I could copy the software to any of my computers or my family's computers. I had already given up on Microsoft Windows and switched to using Macs. Suddenly, I no longer needed to depend on Microsoft Word, Excel, and PowerPoint, which would have cost around $500.

OpenOffice.org ran on my Mac and PowerBook. Not a piece of Microsoft code in sight. And yet, if a client sent us a Word doc or PowerPoint file, we were ready for it. The OpenOffice.org package included the source code, too. Not that we knew what to do with source code, but it was comforting to know that nothing in the application was hidden, nothing was secret. Thousands of programmers had already pored over the source code looking for bugs.

While everyone knew that most software products sucked, according to an anecdote of that time Microsoft products sucked a bit more. The running joke was that Microsoft software was never truly ready for the public until version 3; that is, no matter what the program, don't buy it until version 3. Windows versions 1 and 2 were unstable; versions of Word and Excel earlier than version 3 were worthless. Other software companies couldn't get away with this business practice of first shipping the product, and then updating it later in response to customer complaints, but Microsoft virtually ran the industry at that time with its near-monopoly power.

We didn't expect people to quickly abandon Microsoft Windows or Microsoft Word. Yet it was happening — one desktop, one laptop, one notebook at a time. By 2005, over a million Windows users with PCs purchased Macs to replace them. Like it or not, for some the decision to move away from Windows was political — a reaction to Microsoft's power and arrogance. But for many, the decision was a reaction to the epidemic of viruses and malware that targeted Microsoft software. The Mac was safer for many reasons, but an important one to us was diversity. It was simply not healthy for the industry to have everyone using the same operating system and application software.

The web provided another solution in the form of browsing and searching online. As content moved online, the tools to create the content broadened in functions, and the system you used to create the content ceased to be a factor. Web servers that delivered the content ran on Linux or Unix, content authoring tools that spit out web pages, such as Adobe Dreamweaver and BBEdit, were more useful than Word, and every system had its own browser. Eventually the content authoring tools migrated to services online, such as WordPress for crafting and then publishing blogs, and wiki software for building subject-specific equivalents of Wikipedia.

An industry dazed and confused

More talk of "revolution" filled the air at the prestigious Spotlight conference of 1997, hosted by veteran technology industry journalist Denise Caruso. The theme of "light, not heat" was rivaled in ambiguity with the closing remarks of Barry Diller, "the mind is prepared, but the mind is confused." The theme might as well have been written as a business plan on a cocktail napkin.

The market researchers and company CEOs talked on panels about advertising and other methods that may work someday for making money with content-rich websites. The breezy advice of one market researcher was to "go with the flow" as if these marketing experiments were more like acid trips. Indeed, the market research studies at that time were surreal, and to paraphrase Barry Diller, prepared but confused. One study presented evidence that men, when asked what they find most fun about using the internet, responded with "browsing" and looking for information, while women, when asked the same question, responded with "communicating" (that is, with other people). What did they dislike? Men thought the information should be better; women wanted better navigation. Seems like common sense. Let us save you the research bucks: when lost, men won't ask for directions but will consult a map, while women would rather ask someone.

The gold hiding in this avalanche of nonsense was the notion that websites could build virtual communities around certain types of content, or even certain topics, such as spousal abuse or women's issues or the Grateful Dead. These communities could feed off their members' interests, and sharp entrepreneurs could focus these communities and grow profitable businesses.

The Spotlight conference shined a light on something that was coming, but that something wasn't exactly a money machine. The conference offered a glimmer of hope with examples of working virtual communities where people actually benefited from the experience, various women-issues chat sites, educational sites (especially ones for small children), game-playing networks, and so on — communities where people could exchange views or research links and play games with each other. Some of this community-building could also be directed toward improving the quality of life on this planet. Or so we hoped.

The very public failures of the "Web 1.0" content-oriented sites demonstrated that the cost of creating original, timely content was much, much, *much* higher than advertising and sponsorship revenues. The business model still looked and felt like a magazine, with huge startup promotion costs (transferred from direct mail to the web, but still high). However, cash flow was nonexistent. A website needed to build a major attraction and hope the eyeballs would show up, then sell the eyeballs. But any dangerous looney could start a website and keep it going. The artists and deep thinkers who might have had the potential to reach a large, interested audience were begging for spare change and getting lost in the noise.

Hiding in plain sight, the dangerous loonies were cooking up websites for selling illegal drugs, trafficking in child porn, counting murders, revealing the secret lives of bridge-jumping suicides, recording the last screams of plane-crash victims, and so on. Out of a "deep web" of content not indexed by standard web search-engine programs, a "dark web" emerged to conduct communications and illegal business anonymously, such as trading stolen credit cards, false identities, and passwords. Browsing the web turned out to be as dangerous as navigating Central Park at midnight.

Net scraping

The plan was to document the web and save us all from drowning in perdition. The official name of the project was NetGuide. (As in TV Guide, get it?) Its purpose was to illuminate pathways of greatness through this bubbling and oozing craziness known as cyberspace.

Working with CMP Media, Fred Davis, a veteran tech editor and co-author on our *Adobe Illustrator Handbook for Designers*, led a team that developed this consumer portal and internet search site in August 1996. With Dan Ruby, former editor of MacWeek, I worked on creating bite-sized blurbs about websites (as TV Guide used to create for television shows). Fred and Dan hired a roomful of San Francisco techno-hipsters to work at the edge of the Tenderloin next door to Hastings Law College, writing blurbs and ranking more than 50,000 websites manually. About 200 people were employed on the project, including about 100 freelance writers.

The site included multimedia features and games such as *Where's Barlow?* based on the travels of John Perry Barlow. Many of these features were developed by Marc Canter, whom you met in the previous chapter, and David Biedny, a pioneering digital artist, master of special effects, and author of the first book about PhotoShop. NetGuide was one of the first major consumer web projects to use Java. One of the programmers on the NetGuide project was Craig Newmark, who later went on to start Craigslist. But NetGuide's secret weapon was software that would "scrape" the web to find and rank websites automatically.

When the CMP Media brass from the East Coast showed up to find out why NetGuide Live was earning such low revenue from advertising, the team assembled in the Golden Gate Theater to show off. Canter's wild presentation included animated advertisements (this was *five years* before the technologies would appear), collections of personal profiles known as "social graphs" (*ten years* before Facebook), and communication apps that would enable people to share their images (*fifteen years* before Instagram). The CMP chiefs were amazed, befuddled, and bewildered, and eventually caved into their fears and pulled the plug on NetGuide Live.

Google started operations a year later with its own software that could "scrape" the web to find and rank websites automatically. We all know what happened with that company.

Looking for a good media bar

While Marc Canter's presentation in 1997 to the CMP Media brass presaged many of the features of the coming multimedia web, there simply wasn't enough bandwidth for ordinary people at home to experience it. Multimedia content had slowed traffic on the internet down to a crawl as bad as the Santa Monica Freeway at 5 p.m.

Marc Canter's MediaBand, one of the first interactive music performers.

Perceiving so much promise in web technology and so few ways for individual consumers to experience it, Canter decided to bridge the gap with a multimedia showcase that would demonstrate how venues such as stadiums, concert halls, restaurants, and clubs could present multimedia content to their patrons.

The MediaBar evolved from Canter's MediaBand performances in much the same way that clubs are founded to give local talent a place to perform. The MediaBand was a traveling rave scene featuring cool interactive experiences between audience members and the band. The roadshow helped test ideas and build support for what Canter called the MediaBar, a future pop-up installation in clubs, concert halls, restaurants, and theme parks. Canter turned his house in the Potrero district of San Francisco, just a stone's throw from the SoMa district which was the heart and soul of multimedia development, into a prototype MediaBar, and it quickly became the coolest place in town.

The MediaBar needed documentation to promote the concept. I worked entirely in HTML, incorporating graphics and thumbnail-sized video clips to describe what a typical MediaBar club would look like and feel.

Outside, on your way in, a public video wall would show you the interior of the club, with people at MediaBar stations, in booths, and in the VR arcade. You would take your picture with a digital camera, enter your name, and receive a magnetic-striped card — all new technologies at that time. You could also choose from categories of original cartoon faces, all designed by local artists, divided into menus labeled superheroes, rock 'n' rollers, rare fish, goofy farm animals, fierce-looking beasts of the jungle, generic-looking humans of different races and colors, and dinosaurs. The caricatures menu offered cartoon faces of dead presidents, rock stars, and computer nerds. The club would allow only one use of a special caricature per night, so that there wouldn't be a room full of Elvis impersonators (the exception is the caricature of Pee Wee Herman, for which there was an unlimited supply).

You would make reservations for MediaBar stations by flying through a 3-D VRML (Virtual Reality Modeling Language) simulation of the inside space of the club to select a station or booth. Then off you'd go into the noisy club with your MediaBar smart card, to get a free drink from Apple if you visited the company's website, and a free mouse pad if you tried out Apple's new animation tool on your face. You also would get a special discount on the Tokyo theme room.

Giant video monitors occupy the sides of the room, playing clips from a video library on request. Computer stations are everywhere, playing games, videos, and browsers. Computer memorabilia hangs from the ceiling, and attendants in zoot suits with blinking ties scurry about helping people get settled with their station's controls. Every person logged into a station is represented on the Face Wall, a giant bank of video monitors off to the side of the main stage. A virtual Face Wall appears on the display in Knock, Knock and other applications as a shared resource, and it is available on the web for outsiders to see who's in the club.

Behind the band, a wall of video monitors displays a single picture, a video clip of double-decker buses entering and leaving a terminal, superimposed over a 3D image of a circuit board up close, pilfered from an Intel commercial. Cruise over to the shrine encasing the original wire-wrapped circuit boards of the first Macintosh (from Andy Hertzfeld's collection) and the other memorabilia from the optimistic days of the personal computer era. Next to the exhibit, a Japanese-style beverage machine offers a choice of Sam Adams beer, Mendocino chardonnay, Coke, Diet Coke, Sprite, or Japanese beverages such as Pocari Sweat. (This is years before Red Bull and smart drinks.)

At a computer station your face appears on the Preferences menu and in applications such as Knock, Knock, a "video phone" system in each computer. The station's monitor displays five video windows, and you can elevate any one of the five to full screen. You can track your transactions in the club, and create a personal Party Card for the evening. You can enable your Party Card to show publicly, so that it would appear when anyone clicks on your Face Wall image or identity, even to folks outside the club on the web. The Party Card is a quick way to introduce yourself to a stranger, or to have strangers browse through your interests. You can even put links on the Party Card for your email address and website.

An image of tonight's host, Soledad O'Brien, pops out from behind an icon on the display dressed in a messenger's uniform, announcing a Flying Other Brothers gig at Slim's next month — an advertisement geared to your taste for jam bands, which the MediaBar's Broadcatch technology has already gathered. The station offers access to the web for making any kind of transaction, but it also offers its own custom tickets and reservation system that already uses your preferred electronic payment method for secure transactions, and already knows your seating preferences.

Click the icon for Video Menus (the station offered a slew of icons for various applications, under the control of the Media-Bar's Venue OS), with videos of cuisines from around the world. Tonight's specials are from Jamaica (pork with jerk sauce straight from De Buss in Negril) and Chicago (deep dish pizza with selectable toppings). You can read about the history of Jamaican jerk sauce. An image of tonight's host, in a chef's hat, pops up to remind you that black-eyed peas are on the menu, as well as red beans and rice, which would no doubt appeal to jam-band audiences. Before disappearing, the miniature host recommends Dixie beer, a special tonight from the Deep South. When you submit your order, a progress chart appears, showing the message going through credit approval and so on, all the way to the kitchen. After a minute or so, the chart shows that your meal is on the fire. By this time another zoot suit delivers the beer, and cocktail music starts from the stage.

At your station you try MazeLove, a pseudo-3D environment filled with whacky artistic worlds and gameplay. You can see the face of every club customer currently at a station, mapped onto a 2-D plane representing a simple avatar that can move around and encounter other avatars and objects. As your avatar moves about in MazeLove, other avatars in that same world can see your avatar, and you can see theirs; you can talk to each other, avatar to avatar. You can substitute your live image at the station for your Face Wall image, so that the avatar looks more like a live human being. With MazeLove it's easy to flirt, and easy to get away quickly without apologies.

In the club's VR Arcade, virtual reality experiences are brought in from the most recent SIGGRAPH. You can project your dancing body onto a background such as a landscape, beach, or cliff ledge with Vincent Michael Vincent's Mandala, which uses a blue-screen chroma key and a graphic overlay.

All in all, you have the complete internet experience including video phones, video databases, an audio jukebox, a virtual reality scene with roving avatars, software that tracks your preferences, and a live talk show streamed to your computer, all to give you something to do between sets of the band.

But this was still only 1997. None of this technology existed; iPhones, iTunes, Napster, Second Life, AdClick, Google, Facebook, and streaming video had not yet been invented. The MediaBar was mostly just a figment of Canter's fervid imagination, realized only a few times at events held in Canter's home. Like other wild-eyed entrepreneurs, he was far ahead of his time. There was no way to realize this concept fully — only incrementally over the next decade.

Ten years after the browser

A decade later I walked down a famous street in Los Angeles on my way to the Roosevelt Hotel for the OnHollywood conference hosted by Anthony Perkins, founder of *Upside magazine*. I passed a homeless mental case screaming obscenities at the street, a cackling squad of middle-aged tourists counting the stars on the sidewalk, shopkeepers muttering quietly as they swept up last night's party trash, hustlers peddling dope, hookers peddling sex, and scores of kids texting with their mobile phones, oblivious to their surroundings. Not much had changed in ten years, except for the texting.

During that time, all our documentation efforts involved designing pages for web browsers, using Netscape and Microsoft's Internet Explorer as the two opposing poles of browsing technologies. Essentially it meant developing for one or the other browser until the iPhone arrived, which changed everything — to this day most pages are designed for the iPhone screen and

Apple Safari browser. The documentation is still something you would *read*, albeit on the screen.

It was all so Web 1.0. Online PDF manuals, online "help" systems, and *web portals*, which were web pages organized for easy navigation and search over a limited set of content. We used them to distribute collected marketing content online, using the standards developed in the desktop publishing years such as PDFs, TIFF scanned images, and JPEG photos, combined with MP3 for audio clips and MPEG for video clips. Our clients could set them up internally for their sales and marketing people to login and download whatever they needed to make the sale. I had set up such portals, including customer success stories, for BEA Systems, which was eventually swallowed by Oracle.

The web browser was now the center of the publishing world. During that time I worked with a small company to create games you could play inside the browser window. I had also worked with John Sculley at Live Picture, serving up high-resolution zoomable images inside browsers for e-commerce sites. Live Picture doomed itself by not going public at the appropriate time, due to the embarrassment that its biggest customer was the online version of *Playboy* magazine. Zoomable images indeed! But Sculley also taught me to use single-word slides in presentations, or sometimes just the word and a period — copying the style that Steve Jobs perfected.

Jobs himself was now back at the helm at Apple, promising and delivering an excellent home computer called the iMac, and perfecting his presentation style at the climax of his keynotes by introducing the newest product with the infamous "One more thing" (copied, of course, from Peter Falk's *Columbo* TV character).

At Hollywood and Highland I turned into the Roosevelt's entrance, which was clogged with casually dressed techno folk, mostly from Silicon Valley and San Francisco, mingling with a minority of suited finance types from both coasts and about a handful of Big Media moguls. Inside it was cool, dimly lit, and way too stylish for the streets outside — a virtual representative of old Hollywood. Arianna Huffington and Carson Daly were the most visible representatives of the professional media that recently took the plunge into blogging and video channeling; but the true stars were the geeks and weirdos that had turned professional.

This was a conference that featured, among many demonstrations of video delivery methods, the newly minted MySpace and Facebook social networks, the nascent YouTube and video blogging, and a panel on whether the new media technologies were killing good writing. The younger generation at that time was no longer reading for entertainment — they were watching videos, creating their own videos, and expressing themselves in instant messages.

Big Media had been stalking teens as far back as Elvis. Those older Boomers were no longer a target for advertising, and at the same time, the younger generations were watching much less TV and much more user-generated content on YouTube and other internet destinations. It seemed ludicrous that a geek like Justin Kan, broadcasting his entire life 24x7 using a webcam strapped to his head, could intimidate Big Media, but Justin was there, at OnHollywood, almost as big as the other new Web 2.0 brands. And marketers were waking to the opportunity to not only target advertising better to specific audiences, but also to track how well the ads were doing — something television networks couldn't provide. Big Media was running out of gas.

What's more, the younger generation of social media nerds were not very sociable, nor did they pay attention to the conference panels. They were more interested in the instant comments from the new global peanut gallery attending the conference online, on computers far away. Everyone laughed not when Carson Daly told a joke about having bad hair on stage for the world to see, but when someone out in the world commented back that

Carson's hair looked fine. There must have been around fifty people in the audience — digital entertainment executives, Hollywood wannabe moguls, media journalists, and bloggers. Only three of us were looking at the four executives up on the stage and paying attention. The others were peering at their smartphones, consumed by their screens, and this was before the iPhone became available.

And yet, just about everything at the conference was interesting, including demonstrations by newly minted brand names like Gracenote, YouTube, and Sling Media. You could start your own video channel with pre-designed themes and easy-to-use video management tools. Each channel could be its own social networking site with private or public profiles, blogs, message boards, onsite messaging, user ratings, favorites, friends lists, comments, and photo albums. You could even add a live talk show to your blog, with live interaction over the internet. Services were announced that could time-shift television content and search by keywords. Even more interesting was the ability to tag television content with keywords, to make content discovery easier. Some services were already putting TV shows on the internet for anyone to watch; the fact that it started with reruns and porn was no surprise, but this trend was seriously catching on, with mobile phones the obvious display for these shows.

The show at the Roosevelt would not have been complete without the appearance of DigiBarn cofounder and roving TV producer Allan Lundell with the newest portable video devices, grabbing folks at random and interviewing them on the spot. Allan has been cranking out streaming video for more than a decade and probably has the largest video archive of high-tech conferences ever known to humans. He eventually became DrFuture of the DrFutureShow on KSCO AM 1080 radio in Santa Cruz, CA, a weekly hosted conversation covering the world of robots, 3D printers, nanotech, and crowdsourced culture shocks that changed reality, as well as spotlighting inventions, disruptions, and trends.

The digital transformation was by now complete, taking into account nearly all forms of expression, from text and images through sound and video to interactive gaming. What had not yet come together was a way to deliver all of this, as a unified experience, to the average user. One aspect of the MediaBar that had intrigued us was our online effort at documentation, a first for that time. We were able to serve customers text, images, audio, and video within the constraints of HTML pages with minimum fuss. Application servers provided a basis for serving up help content, user manuals, and eventually public and private wikis.

As I left the Roosevelt bound for Burbank, the same kids were still texting on the sidewalk, the tourists had moved on to the Chinese Theater, the junkies crowded around the hustlers, more hookers were flaunting more booty, and the homeless mental case was still screaming obscenities at the street. Soon there would be more mental cases, screaming or doing whatever they needed to do on video to get attention on YouTube.

And suddenly, *crowdsourcing* became a thing. *User-generated content* became a thing. Most importantly, *getting attention* became *the* thing, the most important thing.

Disrupting influences

Outsourcing to the crowd for information was not new. The British government, looking beyond the expertise of the professional classes, offered a "Longitude Prize" in 1714 to anyone who could develop a reliable method of calculating the longitude of a vessel while at sea; a clockmaker won it. And in 1879, several extremely smart people called for a re-examination of the entire English language, and asked readers to send them references to everyday and unusual words. This was the birth of the Oxford English Dictionary.

As the world turned giddy with nervousness over what might happen as our digital clocks advanced into the year 2000 (the Y2K problem), two enterprising entrepreneurs, Jimmy Wales and Larry Sanger, started to create an English-language internet encyclopedia. By 2001 Sanger and Wales introduced Wikipedia, which permitted users to not only read pages, but also — unusual at the time — *make changes* to them directly in the browser.

Crowdsourcing, according to two editors at *WIRED*, Jeff Howe and Mark Robinson, represents the act of a company or institution taking a function once performed by employees and outsourcing it to an undefined (and generally large) network of people in the form of an open call.

This form of crowdsourcing took hold and Wikipedia went global. As recently as September 2018, Wikipedia tallied 15.5 billion page views for the month, and received over 117 million monthly unique visitors from the United States alone. You may have noticed by now that we use Wikipedia a lot for our historical links.

Crowdsourcing is part of a general trend of participatory activity on the internet known at that time as Web 2.0. Wikipedia is an excellent example of Web 2.0, and its definition of the term defines the entire industry:

> *Web 2.0 (also known as participative (or participatory) web and social web) refers to websites that emphasize user-generated content, ease of use, participatory culture and interoperability...*

As crowdsourcing grew to be generally acceptable for learning companies such as Quora, and for charity funding such as GoFundMe, it also made user-generated content acceptable, and to some even preferable to the professional media. It was the logical extension to desktop publishing, enabling a sort of citizen's media that advocated the democratization of content production and the flattening of the traditional mainstream media hierarchy of prime-time networks, newspapers, and cable news.

Disrupted first were magazines that depended on advertising revenue tied to editorial product reviews, because firms as big as Amazon and as new as Yelp were enabling user-generated reviews. TripAdvisor and a variety of booking sites disrupted the traditional travel agencies. Amazon created an entirely new e-commerce industry that disrupted the big box stores.

As free online services like craigslist killed classified advertising, many newspapers did not survive the transition to online publishing. Ad revenue disappeared entirely, and many online news sites either died or switched to subscription-only services. The death of so many newspapers created countless "news deserts", areas with no local news sources. It is estimated that U.S. newspapers continue to die at the rate of two each week. Local newspapers fold while large-circulation papers are swallowed up by Alden Global Capital, a secretive hedge fund. In fact, hedge funds and other financial firms control half of the daily newspapers in the U.S. They strip-mined the industry by reducing the number of journalists.

Professional pundits understood what was happening. According to our friend Dana Blankenhorn, who publishes the Facing the Future substack of articles promoted by email, "There is no such thing as a newspaper, a magazine, a TV news channel or even a news website anymore. There is only the Web. If you want to live there, you must build a community within it."

Blips in our blogs

As amateur pundits were threatening to crowd out all other voices in the general media, technical writers were losing ground to "user-generated" instructional content on YouTube. As tech columnists, we discovered to our chagrin that similar disruptions were occurring with computer-focused magazines, which had been one source of revenue for us. Other dwindling sources included putting together books and web content, and producing instructional videos.

In 1996, with Cheryl filling in the gaps in our cash flow by switching to teaching, I started a personal online journal that, I hoped, the pioneers of the internet would read. I had some experience with this. In college, a typewriter was my writing instrument, and I was something of a journal writer, which today you would call a blogger.

Tony in college journaling about road trips and Jack Kerouac.

My journal, *The Trip Thus Far*, chronicled my antics in college and extended into real life in the Data General years, eventually covering our early partnership and adventures heading out west. It was posted in the true sense of the word: by regular mail. I don't remember how many issues I wrote and posted, as I have none of them now.

My first mistake was that I couldn't settle on a title for this journal. It started as *Learn to Duck* (in homage to my bandmates in the Graceful Duck). This is how it opened:

This is a periodically updated column about the interactive content development industry from my point of view. Sometimes I rant and rave here before refining the process and producing a column for some other publication; other times, my ravings are not refined but stay right here, just hanging out in the ether.

Retitled *Media Lib* as of the second installment, and finally as the *Bove Report*, it was essentially an online column written as best I could in the gonzo style of Hunter S. Thompson. In many ways it was a blog. The difference was that the column titles appeared in chronological order like a book in the website's menu, rather than in reverse-chronological order, which became the format for blog entries. Not many people read it, because I had no way of promoting it, except by making guest appearances on other websites in the form of links.

Eventually, *Bove's Blips* was reborn as a blog in WordPress, and an early episode titled "What's the New Mary Jane?" talked about new and spectacular personal technology; notably, the iPod and iTunes software for Windows PCs and Macs, the upcoming iPhone, and the world of accessories. I also explained our years-long lapse from the publishing space:

> I started writing something like a web diary way back in 1996, before the blogosphere existed. The last blog entry in Sept. 2003 told about how my band, the Flying Other Brothers, had started a mini-tour. Like Bob Dylan's infamous Never Ending Tour (which lasted at least five years, give or take a year), our tour seems like it will never stop; unlike Bob's tour, ours did not sell out venues and generate cash rewards. Like touring, blogging can also seem like an endless chore with no return. I suffered Blog Depression way back in 1997, and posted less and less over time until I gave it up entirely. Until now.

Amidst entries like Notes from the Liberty Front ("The French government is saying no to Microsoft...") and Don't Play Vista For Me ("You can have a perfectly secure PC if you keep up with the patches. Good luck with that!") I wrote A Song for the Apple Store Line ("Hey! We got to have iPhones today!"). Years later I was still stirring up debate among my friends with Shut Up and Play Your iPhone, Volume 1 and Google Glass Makes Us Look Ridiculous.

Writing and publishing a blog can be fun, as a hobby. As a business, it was (and still is) mostly a failure, unless you are using it as an inexpensive promotional vehicle, such as promoting books that people can buy. Advertising revenue is fickle at best. To promote the blog, we used MySpace posts and eventually Twitter tweets and Facebook posts. After a decade of blogging, all of the action has moved to X (formerly Twitter), Instagram, Facebook, Medium, and SubStack.

Tips in our phones

The blogs we produced for software developer clients were designed to highlight new features of their products, and nearly always provided links to online documentation published in corporate wiki systems. Our Eureka moment was the discovery that the engine behind Wikipedia was available to use online, and a client for it was already available on the iPhone as an app. One could use a wiki to deliver private, as well as public, documentation of our own choosing. There had never before been a medium for instructions and reference content that was always up-to-date, easy to search, and organized for quick reading.

And so, Tony's Tips was born. More effective than an e-book, *Tony's Tips for iPhone Users* delivered continually refreshed and searchable iPhone tips in a convenient app that could also collect user feedback. It provided helpful tips for using your iPhone, such as battery tips, synchronizing with iTunes, tricks with the iPhone keyboard, and setting up email. We were able to make the content available immediately after Apple released the newest version of the operating system.

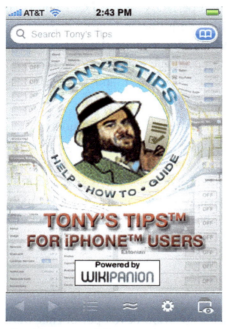

The iPhone app Tony's Tips for iPhone Users.

Our partner on this project, Robert Chin, had developed an iPhone wiki client app for Wikipedia, and we used its engine for *Tony's Tips*. The online wiki content could be updated at any time without affecting the iPhone client. As a result, readers didn't have to wait for a book, e-book, or app update — they could access the refreshed content with the current version of the app. Readers could also save pages to read offline.

Released on Jan. 5, 2009, the app immediately garnered praise from its users. "Tony's Tips should be the first app a new user installs on their phone," said one reviewer in the App Store. "You'll never need a how-to book again," said another. "Within a few seconds I learned a trick... I consider myself an expert user, so I am surprised... and delighted!"

We thought of *Tony's Tips* as an important first step in establishing a documentation system and format for handheld devices. Authors could create and host the content, sell the clients directly to readers, establish direct feedback loops with their readers, and continually update the products easily so that the content is never out-of-date. The client would work indefinitely without upgrades, accessing continually refreshed content.

And yet, Tony's Tips as a business was another failure. Customers resisted spending $2 when so many iPhone apps were free. The death knell came when Apple published its own "help" app for free. We had learned another lesson in high tech: Don't step on the toes of giants. We had embarrassed Apple by demonstrating the need for a helper app, and we paid the price.

The end of documentation as we know it

In the past we loaded our applications into our computers, and then followed the help guidelines or read the supplied documentation (often in PDF). The applications grew into behemoths that could be updated only once a year. The documentation for these applications took on many new forms including book chapters, help messages, animations, and videos.

When applications grew too big, technology came to the rescue with web apps stored in the cloud that could be updated every day. A *cloud* is a global network of servers that can provide computing services over the internet.

Adobe, one of the founders of the desktop publishing industry, consolidated its position as the hub of creative expression by acquiring companies such as Aldus and Macromedia, and packaging their apps as online services. Even the tried-and-true Photoshop behemoth evolved into online services by paid subscription. Google developed free services for consumers such as Gmail and Google Docs, displacing the behemoth apps from Microsoft. As the giants stepped on each other's toes, software development shifted to providing application programming interfaces (APIs) for third parties to develop unique, branded services.

User documentation took the form of free online help, and independent book publishers and authors such as ourselves took another hit. Our *iPod and iTunes For Dummies* (Wiley) lasted a decade, with 11 editions, but revenues from that book and related books in the channel were dwindling. Pivoting into video with Apple's iMovie, we produced several video tutorials about developing iPhone and iPad apps in Xcode.

Video training for iOS application developers.

Even as the need for user documentation dwindled, the need for API documentation grew, and writers such as ourselves adapted to becoming organizers of large quantities of reference information. With API document generation tools, engineers could annotate the code in their source code comments, and writers could edit the comments and automatically generate the documentation. Even better, the Java-based and RESTful APIs could be used directly, accompanied by "Try-It" buttons that let you exercise the API with your own data.

Why read the documentation for the API when you can just click a button to try it?

One company I worked for during this period provided an API for large corporations and business travel services to coordinate employee travel, including booking flights, hotels, and rental cars. Eventually, smartphone applications were developed to use this API on-the-fly. The release notes for each monthly release of the software were many pages long, but I generated them automatically from Atlassian Jira bug reports and responses, integrating Jira with the Confluence Wiki to publish the release notes.

As more publishing moved online, subtle changes were occurring in the culture of the workforce, resulting in a looser environment. Some changes, such as free pizza, were a response to the nonstop hours that so many engineers put in to make a product. Other changes included adjusting to their comings and goings at odd hours and working remotely. Customers and partners in suits and ties were rarely seen in the offices because they could teleconference into meetings and get training online. Even the salespeople wore jeans and T-shirts. Casual attire, online games, and headphones demonstrated how entertainment was pervasive throughout the workplace. It was not unusual for employees to gather together in a common room and play music as well as ping pong and arcade games during working hours.

Once, as the CEO of the travel software company directed some of the board members to the conference room, they had to pass the cafeteria in which some of us were getting down on the blues with electric instruments. To the CEO's chagrin, one of the board members muttered, "I wish I could work here."

"Aren't they supposed to be working?" Tony rocks out with colleagues at a travel software company meeting.

Companies paid good money for writers who knew how to put together API documentation. For decades I worked on these types of online API references for clients very large and very small. Even now as I write this book, I'm working in Markdown and using Sphinx to produce online Python-based API documentation for one of the largest companies in the world.

Good thing I have this job, too, because, as you will see in the final chapters, it looks like the world of media will come to an end soon, with musicians, writers, and other artists unable to make a living, and APIs for artificial intelligence taking over just about everything.

Tools for publishing online

The following are tools we used for publishing online digital media.

Media: Websites, digital books, blogs, online entertainment, games

Tools and languages:

HTML and JavaScript

OpenOffice, NeoOffice, Netscape Navigator, Internet Explorer

Online search, RoboHelp, Tiki Wiki, MediaWiki, Confluence Wiki, Atlassian Jira, database publishing, API generation tools (Doxygen, Sphinx)

Mixing Music Playlists

As modern people try to locate themselves in a world that is changing with bewildering speed, they find music especially rewarding, for music is among the most tenacious cultural elements. Music symbolizes a people's way of life; it represents a distillation of cultural style. For many, music is *a way of life.*

— Jeff Todd Titon, *Worlds of Music*

Zymurgy's First Law of Evolving System Dynamics: Once you open a can of worms, the only way to re-can them is to use a larger can. — Murphy's Computer Law

Nearly all music streams to your portable device or computer for free. But it's difficult to find a particular song if you don't know its title, the artist name, or some other context. Fortunately the streaming services offer *playlists* that help you discover music by organizing it according to context.

As the printed page evolved into a scrollable page for digital text and images, the broadcast radio shows of yore evolved into playlists for navigating digital music. We seek out playlists for love songs, workout songs, driving songs, and whatever is popular among our friends at any moment. Playlists are the easiest way to document a musical mood such as romantic music, a narrative such as a genre or subgenre history, or even music lessons that poke around in a musician's influences. A playlist is the easiest way to organize the music you love to play over and over.

Creating your own playlist is one aspect of music's "desktop publishing moment" — the moment when people took the means of production into their own hands and distributed music without the "help" of record company promotions and radio stations. Creating music without their help was another aspect of the new "desktop music".

Before these advances in media technology, musicians needed studios and producers to record their creations, record companies to manufacture copies, and radio DJs to draw in customers for the copies. This business model led to flagrant abuses and outright rip-offs by the producers and record companies and illegal payola for the radio DJs.

> ### DJs on the air
>
> The term "playlist" originally meant a list of recordings to be played on the air by a radio station. The first use of "playlist" occurred in 1972, used to describe the curations of radio disc jockeys (DJs). The term "disc jockey" was coined by Walter Winchell in 1935.

Musicians have been able to create professional-quality music in garages and living rooms over the last three decades. Consumers didn't have to accept whatever playlists were broadcast, or "published" as the song sequence on an album side. With the introduction of tape cassettes, boomboxes, and the original Sony Walkman cassette player, people could act like their own radio stations and publish their own "mix" tapes.

A "mix tape" cassette (top) and Spotify playlists

As with desktop publishing, the people took control of music creation and distribution in the face of an extremely truculent music industry that fought each innovation by imposing fees.

The record companies with their A&R (Artist & Repertoire) managers and agents had to be dragged kicking and screaming into the new age, and still they profited at every turn, mostly at the expense of the consumers and musicians. Many of us from the Who's "My Generation" remember paying for the same music over and over. We bought some albums, such as the Beatles' *Sgt. Pepper*, six times: A vinyl record, an 8-track tape, yet another record when the first one was scratched, a cassette, a CD, and a download. According to the entertainment industry, even if we paid for a song six times over, we never bought the song, just the media it traveled in.

This perverse logic benefitted the entertainment industry. It was time for a revolution in how to select and play the song of your choice, at home *and* on the road, without having to pay multiple times for it.

On the losing end

Authors' disclaimer: While we appreciate all music, we are partial to the classic rock period of the 1950s-1970s. In particular, we are drawn to documentaries and playlists that provide a historical context for the music.

Many of us who remember those good old days of music devotion don't share the enthusiasm of the younger generations for the convenience of streaming music. We feel that we have lost several key aspects of the music experience in the transition.

One thing we've lost is *tangible value*.

You could hold a vinyl record or CD in your hands, soak in its cover art, and read its liner notes. In the past, albums generated hype, creating an event around the release of a set of songs. A new vinyl record or a new CD was a cherished prize. Today, listeners can cherry-pick individual songs, so artists tend to release singles to streaming audiences without waiting to compile an album. People can hear the music without buying it, so the musicians get paid only a tiny fraction of what they used to get for their work. Loss of tangible value leads to loss in revenues and a dying industry.

Another loss is a *unifying experience* that can boost superstars into the realm of legacy heroes.

There are too many superstars, but not many (if any) rise to the level of prominence and stature as Elvis Presley, Louis Armstrong, Frank Sinatra, the Beatles, the Rolling Stones, or

Bob Dylan. Today there are so many previously-programmed niches representing audience tastes that few listeners have time to venture beyond their comfort zones.

A third aspect we've lost is *trust in the integrity of our influencers*.

Today's influencers on Facebook, Instagram, and X are paid sponsors with no objectivity, just products to push. Music listeners get stuck in their own bubbles, hearing what they already like to hear, and rarely does anything breathtakingly new come through.

The lack of these aspects leads to a *loss of a meaningful context* for the music.

The loss of meaningful context leads to misconceptions and ignorance about the song. An example is how a U.S. presidential nominee at a conference used a Bob Dylan lyric from "Masters of War" without having a clue as to its meaning.

The artists of the late 20th century were catalysts of political change and cultural identity. We can immediately grab the significance of songs that evolved from times of slavery, from the blues, from the Appalachian mountain country, from New Orleans jazz, from the civil rights movement, from the psychedelic experience, and so on.

It's not that historical context is a precondition for good music, but some clue about the motivation behind the songwriter would boost recognition and satisfy the discerning listeners. We want to know *why* the artist creates and performs, what purpose the artist thinks it serves, and under what circumstances did the artist get the inspiration. Today, the best way to find context for music, especially historical context, is to look for playlists that open up vistas in the past and present.

Revolutions per minute

John Lennon of the Beatles, sprouting a mustache and sporting granny glasses, flashed a smile at the cameraman from inside his paisley-painted Rolls Royce. What impressed me the most, as I watched that newsreel back in 1967, was the custom *record player* in the back seat. He could play his favorite phonograph records while commuting to and from London. How wondrous!

Bouncing the stylus

A record player uses a turntable to spin a disc called a phonograph record at a constant speed, with a stylus that slides along a groove in the disc, picks up the sound, and sends the sound to an amplifier and speaker. Phonograph records are seven-inch, ten-inch, and twelve-inch discs made originally from shellac and eventually from polyvinyl chloride; hence the use of the term "vinyl". A bouncing stylus would ruin the disc, which is the source of another familiar phrase in response to someone repeating himself, "You sound like a broken record".

But Lennon's player most likely didn't work very well when the car was moving and its tone arm and stylus with the needle cartridge started bouncing. The typical record player of the 1960s would skip over a portion of a song if there were any external vibrations.

A phonograph LP record in a plastic inner sleeve outside of its cover, and a stereo turntable with speakers popular today.

Our ancestors played ten-inch records at 78 revolutions per minute (RPM). The first 78s at the beginning of the 20th Century were used to deliver classical music at first, and then broadened to provide "pop" music, which was eventually classified for different markets with the "jazz", "blues", and "country" labels. The jukebox, introduced in 1927, was the primary and most convenient way for people to select the phonograph records they wanted to hear and share with others, especially newly released music.

By the 1950s record players were more affordable, and pop music was distributed on seven-inch discs played at 45 RPM that could hold a bit more than three minutes of music on each side. That capacity set the time limit on pop songs to be no longer than three minutes, and the two-sided discs spawned the concepts of the "A" side as the most popular song or "single", and the "B" side as the less important song. For example, I remember my first 45-rpm record, given to me as a gift at age 4, was "The Purple People Eater" by Sheb Wooley, and my second was "Itsy Bitsy Teenie Weenie Yellow Polkadot Bikini" by Brian Hyland. Those songs were the "A" sides; I don't remember the "B" sides. My older brothers collected 45s of Elvis Presley, Chuck Berry, and all the early rock 'n' rollers. From 1950 through 1965, 45 singles dominated the music charts, which not only measured the record's success in sales but also determined whether or not the record would be played repeatedly on radio.

The larger twelve-inch long-playing disc (LP), which played at 33 1/3 RPM, could hold around 22 minutes of music on each side. The LP gave birth to the "album" concept, in which the songs could be related or songs as long as symphonies could be delivered on the two sides. LPs were the first, starting in 1958, to provide stereophonic sound (stereo), which is still the most common format for recorded music.

An amazing innovation at that time was an automatic disc changer for the turntable. You could then stack records on the changer and play them in order. With disc-changers and albums, we had the rudimentary form of the playlist.

Another innovation with LPs was a new form of documentation called *liner notes*, which started out as program notes for concerts and evolved into content such as essays, instrument credits, and even poetry, that were printed on the back of or inside the sleeve that held the LP record. Liner notes often contained a mix of factual and anecdotal details that provided histor-ical and social contexts for the album's musicians.

Lyrics appeared on the back of the Beatles Sgt. Pepper *album.*

Lyrics began to appear in liner notes in 1967, on the Beatles *Sgt. Pepper* album — yet another innovation by the most impor-tant pop group of the 20th Century. (It was Paul McCartney who pushed the record company to make the album a complete enter-tainment package with a lavish gatefold sleeve, quirky cut-outs and badges, and the lyrics printed on the back of the sleeve.)

> ### *Liner notes provide context*
>
> Liner notes sometimes provide metadata that can help when cataloging sound recordings. In particular, we used the label information, copyright date, and so on, when listing used vinyl records for sale on eBay.

LP record sleeves, especially the front cover, helped to sell the album. Once you held the LP in your hands at the record store, you had to buy it. The LP gave off a tactile sense of a rich and cultural artifact.

But how did we learn about the music? What drove us to the record stores in the first place? Radio played a key role of promoting music to the masses.

Getting in tune with radio

Once upon a time, radio was spontaneous, informative, and even vital to the national discourse. Orson Welles terrified thousands in 1938 with his radio depiction of the H.G. Wells classic *The War of the Worlds*. My parents would never forget Franklin D. Roosevelt's "fireside chats" in the early 1940s during World War II, or the Senator McCarthy hearings in the 1950s.

Speaking of McCarthy, the rise of demagogues like him was the direct result of radio. No orator holding forth every night to packed houses of 10,000 people could be heard by as many as could hear one local broadcast in a city of more than 100,000 in just one night. Television accelerated this trend, giving rise to even more popular demagogues that knew how to exploit the medium.

But radio was also fun to listen to, and the DJs were nearly as popular as the musicians. Music reached us through radio stations around the country, broadcasting far and wide into the night, influencing thousands of budding musicians. Radio introduced jazz to the public before the public could afford record players. The history of the Grand Ole Opry radio broadcast *is* the history of commercial country music. The same is true of the "King Biscuit Time" whose DJ was blues harmonica master Sonny Boy Williamson, which started in 1941 at the KFFA radio studios in Helena, Arkansas.

Radio found an entirely new audience of teenagers when it broadcast Elvis Presley, Chuck Berry, Little Richard, Jerry Lee Lewis, and the rest of the rock-n-rollers. I can recall listening after midnight on my portable radio in bed with hormones raging, dialing past the schlock on Philadelphia's WHAT and South Jersey's WYBY, tuning between stations to pick up signals skipping across the ozone, broadcasting out of Memphis, Cleveland, and especially DJs like Jack the Cat and Poppa Stoppa in New Orleans, all this music infiltrating the night sky across America, inspiring the "late people" as the DJs called their late-night listeners, the anxious teenagers staying awake to hear dirty lyrics, weird rhythms, and heartbreak songs.

Hy Lit, DJ at WIBG in Philadelphia, would exhort his listeners: "Calling all my beats, beards, Buddhist cats, big time spenders, money lenders, teetotalers, elbow benders, hog callers, home run hitters, finger poppin' daddy's, and cool baby sitters! For all my carrot tops, lollipops, and extremely delicate gumdrops! It's Hyski 'O Roonie McVouti 'O Zoot calling, uptown, downtown, crosstown. Here, there, everywhere! Your man with the plan, on the scene with the record machine."

At some point in my youth, I recognized myself in that audience of "late people" and started buying the records that I heard on the radio. Before I ever heard the term "frequency modulation" I was listening to album-oriented rock (AOR) music elbowing into the mostly classical music on the FM broadcasting stations. Radios started to have an FM band included with the AM band in the late 1950s and 1960s; its inventor Edwin Armstrong was rendered penniless by the large media corporations of the 1930s and 1940s, and he eventually committed suicide (see *The Tragic Birth of FM Radio*). While the AM band was dedicated to pop music and talk shows, it started to cover more rock music in the early 1970s due to the influence of FM. For example, in the summer of 1971, the AM station WBIG announced a special free Leon Russell concert to occur somewhere in the Philadelphia area. You had to keep listening all day to find out where, exactly, the concert was supposed to be.

Cars had built-in radios. While in high school I would borrow my mother's car and drive out to the edges of the suburbs of Philadelphia, on what we called "the space roads" that led to our primeval psychedelic wonderlands of the Tyler Arboretum, Longwood Gardens, and Valley Forge. FM radio provided the soundtrack for these explorations.

I had dreams of becoming a DJ, but they were shattered in college as I tried to compete with others for time slots on the college radio station. I would help Andy, a DJ friend, come up with interesting music, such as album tracks often ignored by the mainstream, and songs that demonstrated historical influences. But the goal for college DJs was to play the most popular tunes and sound just like the professional stations, in order to get recognized by management and leave college with jobs.

Jeremy Savage, a professional DJ in Worcester, MA in the late 1970s (and in Hartford, CT in the 1980s) explained how songs were chosen *for* him. He had no say in what to play. He was just a voice, a personality on the radio. The songs were picked by a nationwide research firm that based its playlists on listening habits. A DJ named Lee Abrams had pioneered systematic audience research and psychographics, connecting people's lifestyles to their listening habits. He modified the album-oriented rock music (AOR) format from a looser, freeform style to a tighter form, using playlists rather than allowing DJs the freedom to play anything they chose from albums.

The tragedy was that so much of the detailed history of music was lost, and its influence on future artists was not understood. Bob Dylan, the Beatles, the Rolling Stones, Jimi Hendrix, and just about all other great music artists of the 1960s and 1970s were influenced by obscure and eccentric radio, even pirate radio. What they heard in that medium inspired them to create original music.

But the matching of listening habits to content — removing the human DJ from the decision-making process — destroyed radio's creativity and influence on music. Matching content to habits leaves us stuck in our own bubbles, hearing or viewing what we already like to hear and view, and rarely does anything new come through.

In the case of FM radio in the 1970s, nothing new came through unless it was sponsored by a big recording label and designed for a more common taste in music. We had to abandon popular radio stations if we wanted to hear the Grateful Dead, Frank Zappa, rock-jazz fusion, original blues, even folk music. Something had to be done.

Going mobile with tape

Notwithstanding Lennon's mobile record player, there was no easy way to take the music you liked on the road. As its playlists were commercialized, FM radio could no longer be counted on to provide the soundtrack for our car trips or the sharing experience we needed.

The solution turned out to be a form of magnetic tape, which has a fascinating history that includes Nazi Germany, American scientists, hardware manufacturer Ampex, and Bing Crosby. It had been used to record live and studio performances in the mastering process for producing phonograph records, and it enabled radio, which had always been broadcast live, to be recorded for later or repeated airing. Since the early 1950s, magnetic tape has been used with computers to store large quantities of data and is still used for backup purposes.

John Lennon coined the term flanging

Magnetic tape recording, introduced in the early 20th Century, radically altered the making of music at that time, and contributes to this day to the user interfaces — the sliders, faders, and other controls — that we use in software for digital recording. The basic techniques of editing — cutting and splicing to remove, rearrange, or compile pieces of recordings; mixing two or more sound sources into one recording; and fading sound in or out — were all innovations that were made easier with tape.

As engineers reduced the size of tape recorder heads, it became possible to record multiple simultaneous tracks on a single tape. Musicians or singers could be separated into two or three groups, such as recording the rhythm section and the rest of the band separately. If there was a flub on one or the other tracks, it could be re-recorded without re-assembling the whole band.

The tracks were then mixed into stereo tracks (left and right), or one mono track. In the 1950s, most rock and roll records were mono recordings. Stereo rose in prominence during the 1960s. In some early Pink Floyd and Moody Blues albums, the sound seems to fly from the left speaker to the right. A song can be mixed for stereo playback so that you might hear vocals coming from the left speaker and guitars coming from the right, but also drums and bass coming from somewhere in the middle. Stereo speakers can create a field of sound, in which instruments and vocals are balanced in volume across the channels — not set to full volume in one channel. Your brain interprets the audio information as more like a three-dimensional sound panorama.

While audiophiles were using reel-to-reel tape equipment, consumers got their first taste of tape in the form of the 8-track cartridge in the mid-1960s. The Ford Motor Company introduced factory-installed 8-track players as options for the 1966 Mustangs, Thunderbirds, and Lincolns. Tandy sold 8-track recorders and blank cartridges in its Radio Shack stores. By the late 1960s, 8-track decks were preinstalled in all models, and prerecorded releases on 8-track began to arrive within a month of the phonographic record release. Many consumers bought music in this format.

No one knew for a few years, but using tape for an archive proved to be disastrous. While tape was good for short-term use, it was highly prone to disintegration in some environments, and even under the best conditions it began to degrade anyway after 20 years. The audio quality was not as good as an LP record, so many of us bought the music twice: on an LP, and on an 8-track for the car. Fortunately the LP records last a lifetime if you don't scratch their surfaces. I'm still playing the music on records I bought more than 60 years ago.

Before we could get used to the idea of using tape cartridges in automobiles, a more convenient format called the Compact

Cassette arrived, invented by the Dutch company Philips in 1963. Originally designed for recording voice-quality audio for dictation machines, cassettes were developed into a higher quality stereo format by 1971 that outperformed the 8-track cartridges.

From the start we bought blank cassettes in order to record already purchased LP records. During the switch in the early 1960s from monophonic recordings (requiring only one speaker, such as a car radio speaker) to stereophonic, many of us had replaced the typical combo record player and speaker unit with a higher quality solution, using a more sophisticated turntable attached as an input device to an audio receiver, which housed an amplifier for stereo speakers. We could then attach a cassette recorder to the audio receiver and record the records as we played them.

Like many others we knew, we began taping our own playlists — only we called them *mix tapes* — and we used cassettes and either Sony or Nakamichi cassette recorders. The 90-minute format (45 minutes on each side) enabled us to make high-quality recordings of entire albums on each side, such as a Sgt. Pepper/Abbey Road mix tape, with singles at the ends of each side to occupy the precious tape. When recording a two-record jam, such as "Dark Star" into "St. Stephen" into "The Eleven" into "Turn On Your Love Light" on the Grateful Dead's *Live Dead*, we needed to be quick with the pause and record buttons.

At some point we started mixing songs to document a concert or time period, such as "Mardi Gras Madness" and "Dead at the Matrix". We defined historical playlists, such as "Country Blues Roots" and "The Red Album" (a fantasy of what the next double-album from the Beatles would have been after the release of *Let It Be*, using the Beatles members' solo albums for versions of songs). Although bands would try concept albums and fail, fans would do their own versions of concept albums with mix tapes.

Meet us in the taper section

Cassette recorders in the 1970s were small enough to take into concerts. Over 14,500 different live recordings of Grateful Dead concerts were distributed to the world through independent tapers who at first had to hide their equipment at shows, but were eventually encouraged to bring their best gear to make the tapes.

Going mobile with CDs

The benefits of recording music with entirely digital equipment were not immediately obvious. Digital recording of music is a process of converting analog signals of the sound, picked up from a microphone or transducer, into a series of discrete numbers that represents the voltage level or air pressure of the highs and lows of the sounds. The first digital multitrack pop-music recording was Ry Cooder's *Bop Till You Drop* in 1979, released on a vinyl LP. According to Cooder, there's a thinness to the sound of the LP due to digital processing that undermined the performances. But the CD reissue of the album, released many years later, demonstrates the recording benefits of high-definition digital recording. And for that era, the digital album sounds crystal clear with great impact due to stellar engineering.

As you learned previously, CD-ROM discs and drives were introduced around 1988 to store data. The format was based on the original compact disc (CD), a digital optical disc with a data storage format for storing and playing digital audio, co-developed by Philips and Sony and introduced in 1982. Engineers adapted his optical disc format into products including Photo CD for storing photos, CD-i for playing multimedia games, Enhanced Music CD (CD+) for music, and so on. By 1985, CDs were the medium of choice for consumers buying music. and CD-R recordable discs replaced cassettes for recording your own music or copying music from vinyl records.

The digital format stores audio with an unchanging tone quality through time. The digital recording process provides high accuracy for picking up even the subtlest sound as well as an overwhelming bass. Together they could be used to preserve music in the digital format without loss in quality. The only variable is the medium used, which can affect quality.

Unless, of course, the discs fail. Even if they were not damaged by getting tossed around outside their cases, the CD-R discs became unreliable after 10 years, and the commercially released CDs started to fail after 15 years. The big music labels didn't lose sleep over this, because they profited when you bought the same music more than once. They also received a royalty on blank CD-R discs to offset what they considered to be losses from rampant piracy. What *did* cause them to lose sleep turned out to be a device as portable as a handheld radio that carried digital music.

Going mobile with the iPod

What if you could fill up your car with music as easily as filling it up with fuel? Introduced way back in 2001, the Apple iPod was truly innovative for its time. The very first model could hold more than 1,000 typical songs. Like the hugely successful Sony Walkman portable cassette player of the previous decade, you could take it anywhere and listen with headphones.

The iPod was, essentially, a hard drive and a digital music player in one device, and the device was such a thing of beauty and style and so highly recognizable that all Apple needed to do in an advertisement was show it all by itself. The standard model weighed less than two CDs in their jewel cases and could hold roughly 1,000 songs. The iPod mini was smaller than a flip phone, and the iPod shuffle was as much a fashion statement as a state-of-the-art music player — less than an inch wide and about a third of an inch thick, weighing little more than a car key or a pack of gum.

The iPod classic. It could display images as well as play music.

When the iTunes Music Store opened, the iPod became something else again: part of an ingeniously conceived blend of hardware, software, and content that made buying and playing music ridiculously easy. iTunes was originally developed by Jeff Robbin and Bill Kincaid as an MP3 player called SoundJam MP, and released by Casady & Greene in 1999. It was purchased by Apple in 2000 and redesigned and released as iTunes. By all accounts, Apple succeeded in offering the easiest, fastest, and most cost-effective service for buying music online.

And yet, some subtleties and complexities needed to be addressed, such as how to manage your music library on your computer and synchronize playlists with your iPod.

We had proposed a book on the iPod, but it took quite a bit of arm-twisting to get a publisher interested. A year after writing our first book on this subject, *The iPod Companion* (M&L Publishing), which focused on using a PC and MusicMatch Jukebox, we finally convinced Wiley Publishing to let us do *iPod and iTunes For Dummies*. We wrote and edited this book through ten editions in ten years, covering every iPod model and version of iTunes up to 2013.

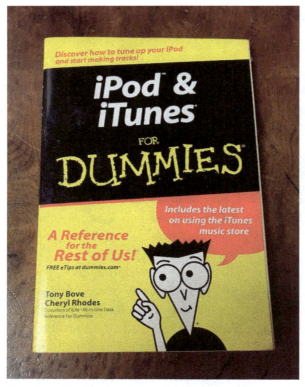

Our first "For Dummies" book, iPod and iTunes For Dummies *(Wiley), was published for over a decade in ten editions.*

Ripping it up

One of the complexities was how to "rip" (import) your music from CDs or even from tapes and vinyl records into your new digital music library. Choosing which encoder and settings to use depended on the type of music, the source of the recording, and other factors, such as whether you planned to copy the songs to an iPod or "burn" (store contents on) an audio or MP3 CD.

Power was also an issue with portable players like the iPod. Playing large files took more power, because the hard drive inside the iPod had to refresh its memory buffers more frequently to process information as the song played. Otherwise you would hear hiccups in the sound.

The audio compression methods that are good at reducing space also have to throw away information, degrading sound quality in the process. In technospeak, these methods are known as lossy (as opposed to lossless) compression algorithms. The AAC and MP3 encoding formats compress the sound and the file size using lossy methods. Lossless encoders, such as Apple's Lossless encoder, compress the information only slightly with no loss in quality or information, but the resulting files are still huge. The AIFF and WAV encoders, also lossless, do not compress the sound at all and are still the best choices for burning CDs, since it doesn't matter how huge the files are.

The encoders provided with iTunes and now with Apple Music offer general quality settings, but you can also change them to your liking. With the MP3 or AAC encoders, the amount of compression depends on the bit rate that you choose and other options. The bit rate determines how many bits (of digital music information) can travel during playback in a given second. Measured in kilobits per second (Kbps), a higher bit rate, such as 320 KB, offers higher quality than a bit rate of 192 KB, because the sound is not compressed as much — which also means the resulting sound file is larger and takes up more space, which in turn means fewer songs can be stored on the device.

We still rip music from vinyl records collected over the last century. It takes a bit of extra time, first to clean the record, and second to record the music as it plays. We use software such as Audacity to remove pops and clicks, which are common issues with vinyl. We also use a noise reduction filter to remove hiss, a common problem with tape.

In the hands of obsessive music enthusiasts and audiophiles, these and other digital audio tools opened up a universe of "bootleg" and live recordings of famous artists and studio outtakes from popular albums — such as the Yellow Dog CDs of Beatles studio outtakes, and the vinyl double-LP "black album" of Beatles outtakes from the *Let It Be* sessions (*not* the Black Album compilation of Beatle solo material by Ethan Hawke).

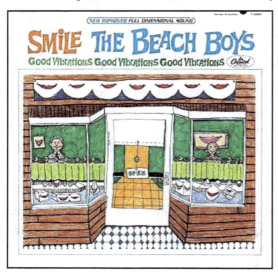

Artwork for the unreleased Smile *album by Brian Wilson and the Beach Boys.*

One example, important to us, was an album by the Beach Boys' songwriter Brian Wilson, recorded as he began to unravel in the post-psychedelic sunset of the late 1960s. The album called *Smile* (or "SMiLE") that Brian Wilson worked on in 1966-67 was never released, but the myth surrounding *Smile* grew to such immense proportions that many, many websites were dedicated to reconstructing it, complete with mixes and edited versions of songs in MP3 format for downloading and creating your own *Smile*.

The backstory is that Brian's previous record, *Pet Sounds*, had been recorded in an atmosphere of friendly but anxious competition between the Beach Boys and the Beatles to be the number-one pop group of 1967. But not just a pop group — to be a *pop art* group. The Beatles were about to release *Sgt. Pepper*, and Paul McCartney has said from time to time that he was directly influenced, especially in his style of bass playing, by Brian Wilson, and specifically the *Pet Sounds* album.

Brian Wilson's *Smile* concept album was far ahead of anything at that time in pop music. In April of 1967, Paul McCartney stopped by the studio in Los Angeles and played Brian a tape of "A Day in the Life" from the as-yet-unreleased *Sgt. Pepper* album (he also contributed some percussion to a *Smile* track called "Vegetables"). The professional competition was very real, the stakes were very high, and the high regard for each other was mutual. There had been nothing like the concept of *Smile* in pop music before this, and *Sgt. Pepper* would set a new standard a few months later, at the start of the Summer of Love. We are talking about a meeting of the minds of the highest order, a pop-music counterculture summit involving the most famous music avatars of the Sixties generation. Just what did happen when Brian heard "A Day in the Life"? Did it cause him to rethink *Smile* and eventually abandon it? Did Paul, after contributing to "Vegetables", really base his style for his upcoming solo album on Brian's inventive harmonies and childlike melodies?

What kept this myth alive was the notion that Brian Wilson was into something incredible, a sort of Holy Grail of psychedelic music, that was stifled by the record company and by the other Boys, who were not aware of the work's artistic merit. The *Smile* album was an abrupt turning point in the history of the Beach Boys, and it was never released. The other Boys did not want to depart from their commercial formula.

An essential ingredient in the mystery of the *Smile* sessions is the perception of Brian Wilson as a misunderstood genius. During this period and after, he withdrew more and more from reality. His frightening experiences with drugs and strange behavior through the 1970s and 1980s seemed to have their roots in the final *Smile* sessions, when the group nearly broke up (some say it actually did, and Brian went along with the myth for years). Armchair psychiatrists pronounced him crackers, while family members tried to protect him. Brian never again commanded the full attention and cooperation of the Beach Boys, and therefore ceased to be able to exert artistic control over the group's output.

And so it was that Brian's pop art masterpiece was lost — until the internet revived it as a major obsession. Websites devoted to the Beach Boys and Brian Wilson began to appear with pages dedicated to information about *Smile*, and eventually, with downloadable tracks for the mythical album. The internet made it possible for fans to discuss every aspect of Brian's mental illness, drug bouts, and emotional frailties as expressed in this music. Due to the powerful myth-making and obsessive forces of the internet, Brian Wilson achieved far more notoriety and critical acclaim for this unreleased, unfinished album than he ever received for his other solo works.

We created our own version of the *Smile* album, culled from tracks found on the Beach Boys box set and from bootleg CDs (see *Surf's Up! The Beach Boys Smile Sessions*). The internet changed the relationship between obsessive fans and artists: now fans can not only be published but can also create mixes and edits of albums.

What's next, a rewrite of the Who's *Tommy*? (Actually, that's been done, in an off-Broadway play, film, and musical.) A new edit of the lost Jimi Hendrix album, *First Rays of the New Rising Sun*?

How about fans re-doing the Beatles' *Get Back* album, the one that was pulled and eventually replaced with *Let It Be*? Why not just reunite the Beatles? Edit tracks from each of their solo albums and create a new Beatles album? (This has also been done – see our own *Beatles Solo Fantasy – The Red Album*).

Could this obsessive behavior change the artist's point of view directly? Brian Wilson responded, not just by putting up his own website, but also by releasing a new version of *Smile*. Nearly 40 years after it was initially conceived, the released version — produced entirely by Brian Wilson with a new band, and spurred on by internet fanaticism — became available in 2004. The unreleased material from the *Smile* sessions in 1966-67 has also been released on Apple Music.

Enthusiasts tried to offer the music directly to the public, but copyright issues closed down the downloading sites. Obviously, the downloading of never-released music was and still is illegal, as either Brian Wilson or Capitol Records (for which the music was recorded) could claim ownership. But that is entirely beside the point. Fans worshipped Brian Wilson and would buy anything the man cared to release. Not even Brian could argue that his reputation had been damaged by the release of crude outtakes — people still treat them like Zen koans from a pop-music master.

The rights stuff

Copyright issues were plaguing the music industry when Apple announced iTunes in 2001. Apple chairman Steve Jobs remarked that other services put forward by the industry tended to treat consumers like criminals. He had a point. Many of the services added a level of copy protection that prevented consumers from burning more than one CD or from using music purchased on other computers or portable players.

Record labels had been dragging their feet for years, experimenting with online sales and taking legal action against online sites, such as Napster, that allowed free downloads and music copying. Although the free music attracted millions of listeners, the services found themselves under legal attack in several countries. The digital music found on these services wasn't of the highest quality, and in some cases, the music was disguised with intentional misspellings in the song information and artist names. Consumers grew wary of file-sharing when the Recording Industry Association of America (RIAA), a trade organization looking out for the interests of record companies, began legal proceedings for copyright infringement against ordinary law-abiding folks who had had the unmitigated audacity to download music.

The entertainment industry has a history of losing the public's trust. It has fought every new development that seemed to threaten current revenue streams. It could not countenance the offering of content through new channels it didn't completely control. It fought the player piano at the turn of the previous century in order to protect sheet music, until piano rolls opened up gigantic new music markets. It fought cassette tapes until record companies figured out how to sell pre-recorded tapes at truck stops. It fought the VCR until corporations figured out how to make billions with a video after-market. Technology companies have always toppled entertainment industry opposition by opening up new opportunities to sell products and services, not by doing the entertainment industry's bidding.

Birth of digital playback

The player piano was the first digital recording and playback system. Publishers copied the notes from sheet music to make the holes (early analogs to digital zeros and ones) in the rolls, which they then sold by the millions. Popular music categories for piano rolls included classical, ragtime, and boogie-woogie.

1920s player pianola (top) and piano rolls.

The fight over copy-protected digital media pitted our two favorite arch-enemies: Microsoft and Apple. Neither company was innocent; both were quite willing to court the entertainment industry's demands for stricter control over copying the music and video you had already purchased. In particular, music labels enlisted technology companies to help fight a losing battle against music enthusiasts who distributed free MP3 songs. Caught in the middle between the music labels and the enthusiasts were the consumers who simply wanted to maintain indestructible backups of their purchases.

Apple was criticized for offering a closed, proprietary platform — the iTunes online music store and the iPod, which was the only type of player that played songs purchased from iTunes. Apple captured nearly the entire market for portable music players and online music stores with iPod models that were irresistible as well as fashionable. Despite its closed nature, the platform accommodated the standard MP3 format as well as its own protected format for purchased music. Apple also let you burn recordable CDs of purchased music, make unlimited backup copies, and copy the music to any number of iPods, but the service still tied playback to a specific set of computers that had to be authorized to play the protected songs.

This tradeoff was still a bit more liberal than the restrictions imposed on purchased music that used Microsoft's protection technology. But no other company, and certainly not Microsoft, was able to match the iTunes and iPod combination for convenience and popularity.

Send in the clones

Microsoft worked hard to establish the Windows Media format for audio and video in the entertainment industry, establishing a broad alliance with Disney in 2004. Competitors to the Apple iTunes online store, such as the improved Napster and Microsoft's MSN Music, sold music in the Windows Media Audio (WMA) format supported by dozens of would-be iPod killers. The iTunes competitors offered all-you-can-hear services that presaged today's music streaming services in that they charged a monthly subscription fee.

However, the iTunes competitors were hampered by a business model that treated songs as "rented" rather than purchased — the songs would disappear from your device after a given period. In early 2005, Microsoft introduced copy protection technology that added a timestamp to WMA files for use with portable music players, so that subscription services such as Napster could offer rented songs that expired after a given amount of time.

For example, Napster to Go made available the company's catalog of one million songs with time-stamped tracks. Every time the music player was connected to the computer, the songs were checked against the user's subscription, and they didn't play if the user hadn't kept up with the payments.

It seems perverse today, but in 2004 Microsoft courted the entertainment industry with copy protection built into Windows for use in PC clones. The Secure Audio Path (SAP) technology forced the certification of device drivers in order to make protection schemes work, thereby limiting consumer choices to hardware that enforced the rules. In a sense, Microsoft was trying to lock users into its chosen hardware the way Apple did with its proprietary hardware. Windows Media Player could obtain information about the songs you ripped from CDs and the DVDs you played, and report that information back to Microsoft so that Microsoft could extend copy protection to your digital versions.

Imagine if the companies that made phonograph turntables in the previous decade were able to decide which records you could hear. Media players that enforce digital rights management made it harder for people to share the content they purchased or to play that content on different machines. It was bad enough that double-clicking a music or video file or inserting a music

CD or a movie DVD into a typical Windows PC would cause Windows Media Player to start up automatically, as if it were the only software on the machine allowed to play it.

The Microsoft copy protection in Player turned out to be even worse. Mercenary hackers discovered ways to subvert it to install spyware, adware, dialers, and computer viruses. When you tried to play an infected media file, it tricked the system into searching the internet for the appropriate license to play the media file, and redirected you to a website that downloaded the malware. How ironic it was that consumers were screwed again: spyware and viruses could be activated by the so-called "anti-piracy" solution.

Buying or "renting" music should be a decision you make because you like the content, not because you are forced into it by your playback equipment. Even the wealthiest artists, who had more to lose from piracy than others, realized that digital media should not be penalized with clumsy protection schemes that backfire by alienating the consumer without stopping the real pirates.

So, in 2007, Apple took a sad song — the copy restrictions on its music — and made it better by removing the copy protection for its high-quality music format. Apple announced this in accordance with a deal with record giant EMI, the record label that still controls a vast database of classic rock from the Beatles, the Rolling Stones, Pink Floyd, and many others. Apple also put together a digital album format for the iTunes Store that took full advantage of album graphics, lyrics, and liner notes.

Apple billboard advertisement for the Beatles on iTunes.

Beatles music was a turning point in the history of music LPs. Beatles CDs were also a turning point in the history of CD sales, and Beatles music once again proved to be a turning point in the history of online music. In 2010, as CD sales plummeted, the iconic Tower Records failed, music downloading increased exponentially, and high tech titans called for an end to copy protection, the Beatles remastered all 13 official albums and released them not just on iTunes but also on the other music downloading services. In just two months, sales had reached 5 million songs and 1 million albums. Paul McCartney praised the deal, saying: "It's fantastic to see the songs we originally released on vinyl receive as much love in the digital world as they did the first time around."

Kick out the jams

Making music is a tradition in every culture on the planet, and serves as a global language that everyone recognizes and understands. Blind Lemon Jefferson wrote songs over a century ago, traveled around the dusty countryside singing and playing the blues, and died penniless, but one of his songs is included in a probe that is heading out of our solar system. The man's music will live on forever.

Primitive societies used brass, animal horn, bone, ivory, even gold — the oldest extant lyre is Sumerian and made of gold, with gold and silver strings. Technology marches on, and instruments change with the times. In the 16th century many "new" instruments were made of wood, and by the 18th century the technologies of woodworking and metalworking made the piano possible and Ludwig van Beethoven inevitable. By the 19th century, Adolphe Sax was so brazen as to combine a wind instrument and a brass horn to invent the instrument that now bears his name, the saxophone. Not surprisingly the technology of electricity, and eventually the microprocessor, would change musical instruments, thereby changing the music itself, forever.

The Moog and ARP synthesizers of the 1960s and 1970s were large, odd-looking and odd-sounding machines based on analog electronics that used electric voltages to create and control sounds. Higher voltages made higher notes and lower voltages made lower notes, and bands such as Emerson Lake & Palmer (EL&P) and Genesis used special keyboards to play them. Keith Emerson of EL&P, and Rick Wakeman of Yes, used extravagant multi-keyboard configurations in which each instrument was set up to produce a single sound per show. Joe Zawinul of Weather Report developed a unique technique for playing on two keyboards simultaneously, placing himself between a pair of ARP 2600 synthesizers, one of which had its keyboard electronically reversed, going from high notes on the left to low notes on the right.

Over time these devices were equipped with programmable memory so that sounds the musicians created earlier could be stored and recalled later for live performances. The layering of sounds upon sounds became an important tool and almost a trademark for many artists.

The next big step came in 1979: New keyboards were equipped with computer interface plugs so that they could be connected to other synthesizers. Development moved swiftly as more companies got into the act. The diversity of keyboards, drum machines, sequencers, and other musical devices grew rapidly. The music technology companies followed the lessons learned from the computer industry and developed a standard for interconnectivity. MIDI stands for Musical Instrument Digital Interface, and is now an international standard that specifies how musical instruments with microprocessors can communicate with other microprocessor-controlled instruments or devices. The first synthesizer to speak MIDI was the Sequential Prophet 600 in 1983, played by some of the greatest keyboard players in jazz and rock.

MIDI communicates performance information, not the actual audio waveform. A MIDI device can register what note you played, how hard you played it (how much pressure was applied to the key of a keyboard), how quickly you released it, and other controls such as sliders, wheels, switches, and pedals. The information is then passed to another device that "plays" the music based on this performance information.

Making music has been part of the Apple Macintosh DNA since day one, when Steve Jobs introduced the original Mac to an audience and used it to play music (simple tones, but it was the first personal computer with built-in sound). Jazz great Herbie Hancock jumped on the Mac bandwagon early, using it to control synthesizers and compose music, as did electronic music godfather Vladimir Ussachevsky and pop/rock icon Todd Rundgren.

The first true program to make music on the Mac was MusicWorks from Hayden Software, written by Jamie (then Jay) Fenton (who went on to create VideoWorks, the foundation for

Macromedia Director). The Mac became the dominant platform in professional music and audio recording, and Mac software has won awards in the music industry; Digidesign's Pro Tools even won an Oscar. Today, Apple's Logic Pro is the most popular music production software for professional studios, and ScoreCloud is one of the popular music notation apps to turn songs into sheet music.

Documenting music with notation

The principal way to document music for creation is sheet music, which uses a musical notation of symbols. The types and methods of musical notation have varied between cultures and throughout history. Even today different styles of music and different cultures use different notation methods and symbols. For example, professional country music session musicians use the Nashville Number System. The most common form of written music used by other session musicians includes a lead sheet that specifies only the melody, lyrics, and harmony, using one staff with chord symbols placed above and lyrics below. A chord chart or "chart" provides basic harmonic information about chord progressions. Some chord charts also show rhythm using slash notation for beats.

Musical notation

Apple's GarageBand, free with every Mac, brought the lofty capabilities inherited from a legacy of innovative music software down to the level of the rest of us who just wanted to make music. As the name implies, you could kick out the jams with GarageBand and record studio-quality music in your garage, home, or wherever. To this day GarageBand offers built-in instruments, special effects, thousands of pre-recorded loops, and the wisdom of at least one or two recording engineers. You can use royalty-free loops in your songs, play the synthesized instruments supplied with GarageBand (and add more from extra instrument packs), and even plug in a real guitar and use GarageBand's built-in amplifier simulators.

In his 34 years as a musician, Pete Sears has played keyboards and bass guitar with a large variety of artists, including Rod Stewart on the classic albums *Gasoline Alley*, *Every Picture Tells a Story*, *Never a Dull Moment*, and *Smiler*. He was with Jefferson Starship from 1974 to 1987, and Hot Tuna after that, and has performed with a who's who of the San Francisco music scene, playing sessions on more than a hundred albums. Sears and I played together when he became the musical director of the Flying Other Brothers in the summer of 2000.

Pete Sears played keyboards and bass guitar with a large variety of artists, including the Flying Other Brothers.

In my book, *The GarageBand Book* (Wiley), Sears had this to say about GarageBand's uses:

> *Composing music is experimental by its nature. If you can experiment with sounds, chord progressions, and rhythm tracks whenever the moment strikes you — a professional musician could be incredibly more productive just in getting sounds and melodies together, and composing bits and pieces, then playing them back. GarageBand is a very powerful tool for doing that — pulling ideas together.*

GarageBand applies MIDI information to anything you specify as a Software Instrument, effectively turning your Mac into a fully functional music synthesizer. It suddenly became supereasy to play music in any location with an Apple laptop and a USB MIDI keyboard.

Music is still a driving force in high-tech culture, just as technology has been a driving force in popular music. Many engineers in the computer industry played music on the side. Like engineers, musicians are especially adept at getting their hands onto something and using it.

I sing and play harmonica, and I never expected to find an iPhone app that would be useful for those purposes. But since 2009 I've been an avid user of Cleartune, a chromatic instrument tuner and pitch pipe that uses the iPhone built-in mic. I can quickly find the proper pitch for singing. On stage, I've used it to quickly show me the root key of a song so that I could grab a harmonica that's in the right key. I also use it to show whether older harmonicas are out of tune and need to be cleaned or adjusted. Others use it to tune acoustic or electric guitars, bass, bowed strings, woodwinds and brass of all sorts, and any other instrument that can sustain a tone.

So, with all this desktop music studio gear, why not start a band?

Playing in the band

What happens when you bring together assorted misfits from Silicon Valley and Palo Alto, devotees of Jerry Garcia here, inventors of Apple II circuits there, and bon vivant sages of the Sixties culture everywhere, into a park across from Peet's Coffee in Menlo Park on a Saturday afternoon? Why, they break out guitars, bongos, and various other instruments, and make music!

The Graceful Duck

I brought out my harmonicas to play
with a duo of Grateful Dead
diehards who played acoustic guitars. We called ourselves the Graceful Duck,
and our short-lived career peaked in a psychedelic afternoon at Stanford University's Student Union in 1981.

The era of the "corporate" bands began to flourish. Professional musicians could get day jobs in high tech, and professional high tech employees could moonlight as musicians. Amateur
bands would play company picnics and Christmas parties. A
loose congregation of jamming high-tech industry heavyweights
played the most influential invite-only conferences of the 1980s
and 1990s, including Stewart Alsop's Agenda and Esther
Dyson's PC Forum.

At one of Dyson's confabs, well-known industry venture
capitalist Roger McNamee had been cut from the invite list for a
high-profile game of charades that featured the stars of the
conference. I was happy that Roger had decided not to fight his
way into this game but had instead joined myself and a few other
lesser luminaries for a jam session in a nearby closet. The band
Random Axes was formed, T-shirts were designed, and touring
jackets were distributed before anyone knew anything about
them.

The jam sessions were unusual, with Silicon Valley's newly
minted billionaires playing alongside pundits and members of
the business press. Everyone remembers when Lotus CEO Jim
Manzi got up to sing "Let It Be" and Microsoft's Paul Allen
cracked open songs with his stinging guitar imitating Jimi
Hendrix. PC Magazine's Bill Machrone crooned the Penguin's

"Earth Angel" and Aldus' Paul Brianerd strummed Lennon's "Working Class Hero". Pink Floyd's Scott Page joined the fun with his sax, Apple's Larry Tesler drummed out the rhythm, and marketing whizz David Szetela percolated the mix with electric guitar.

Two decades later, out of the smoldering wreckage of the San Francisco dot-com economy, the Flying Other Brothers emerged as modern-day Merry Pranksters, conjuring up an intoxicating blend of neo-psychedelic music and seamless ensemble jamming with a large repertoire of original tunes and stunning covers.

Tony singing and playing harmonica with the Flying Other Brothers at the Great American Music Hall in San Francisco, 2002.

Once called "Silicon Valley's favorite four-car garage band" by the *San Jose Mercury News*, the Flying Other Brothers included remnants of Random Axes and other amateurs from high tech, including Roger's brother Giles McNamee, venture capitalist Larry Marcus, high-tech consultant Bill Bennett, Bert Keely, architect of the Microsoft Tablet PC, and myself. We had spent three consecutive summers taking a "band camp" course developed specifically for us by Jefferson Airplane guitarist Jorma Kaukonen, who assembled a faculty that included G.E. Smith of "Saturday Night Live," Jack Casady (Kaukonen's partner in the Airplane and Hot Tuna), and Pete Sears of Jefferson Starship. G.E. and Pete eventually joined us, along with Barry Sless (from the Dave Nelson Band and Phil Lesh) and Jimmy Sanchez (Bonnie Raitt and Boz Scaggs).

The Flying Other Brothers (from left): Pete Sears, Bert Keely, Tony Bove, Roger McNamee, Bill Bennett, Ann McNamee, Jimmy Sanchez, and Barry Sless.

We were a professional band from 1997 through 2006, working more than 80 gigs a year, sharing concert bills with Little Feat, Leftover Salmon, Steve Kimock Band, and String Cheese Incident, and backing Grateful Dead members Bob Weir and Mickey Hart in a series of fund-raisers, and appearing on stages ranging from the urban centers of Alaska to the wilds of New York City. We even won the Rock and Roll Hall of Fame in Cleveland's "Battle of the Corporate Bands" (second place was the band from the Harley-Davidson company).

In the Spring of 2001, as the computer industry took a nosedive and most of us lost our day jobs, the band suddenly took on an entirely new life. The surprise of a lifetime was our ability to book Abbey Road Studios for one night, to lay down a few tracks for our first commercial CD, *52-Week High*. Many a Beatle fanatic will tell you what it's like to stand outside the gates of EMI's Abbey Road Studios in London. I've stood outside those gates on just about every trip I've ever made to the U.K., as if on a pilgrimage to a holy place. I've admired the graffiti on the white walls facing the street, and took photos, of course, of friends crossing the zebra crosswalk that appears on the front of the Beatles Abbey Road LP and CD.

I'd once been the hippie outside, sneaking around trying to add graffiti, and now I was strolling in like a conquering hero, a true musician. The receptionist greeted me warmly, as she would have greeted any band member (the previous band to walk in had been the Small Furry Animals).

Awestruck, completely blitzed by atmosphere, I walked down the long staircase from the control room into studio #2 and took my place next to my fellow Flying Other Brothers. At that time Studio #2 had many of the original Beatles recording microphones and keyboards (including the "Lady Madonna" piano and vocal microphones used by John Lennon, one of which I used for harmonica). The music we recorded in that short session of one night was inspired, sweeter than anything we've ever done. I remember looking up to the control room from below and thinking of the gods of Mount Olympus. How many bands have recorded here, using Beatle equipment, looking up those intimidating stairs to the control room, and hoping that the engineer at those controls understands their music. The Beatles had started out as mere mortals in this very studio, and with the help of the right mix of a visionary producer (Sir George Martin) and his engineers, had risen to the task to become the most inspirational musical group of all time.

We recorded several overdubs for the album and were given a tour of the place, including the infamous bathroom where the Fab Four would retire for serious smoking between sessions, the cavernous studio #1 that once hosted the orchestra for the "A Day in the Life" session, and the gold-inlaid mixing board for Pink Floyd's *Dark Side of the Moon*. The studio personnel were very friendly and experienced at giving tours. The studio takes photos of the graffiti on the outside wall every year before whitewashing the wall, and keeps the photos on a bulletin board for each year.

After getting drunk in the canteen and hanging out in the inner garden on the bottom floor, I realized that Abbey Road was a really nice place to just hang out (if you could get permission to do so). Part museum and part working studio, it has hosted the cream of British rock for half a century. To be there and make music made me feel like I had attained a new level in the game of life, a level that requires new skills and offers new opportunities. A chance to tease the gods of Olympus.

I didn't realize how much the Beatles individually had influenced my life until the passing of George Harrison, just one year after I'd visited Abbey Road. When John Lennon was killed in 1980, I had coincidentally decided to change my career from corporate worker to freelance writer and journalist, and to get back to playing music in a band. In other words, work a day job in order to support the music habit. But as George said in *Yellow Submarine*, "It's all in the mind." And John had always embodied the spirit of activism, of going out and just doing it, and by reviewing his life I found the courage in myself to make the change.

For what it's worth

We applaud the myriad ways that technology has unleashed the creative spark in people to create music. With millions of musicians surfing the planet waves, there's an awful lot of good music out there, some of it great. While many of us pay for and download albums by artists we know about, the most popular method of exploring this universe of unfamiliar music is through streaming services like Apple Music and Spotify — either free (with ads), or by paying a low premium equivalent to buying an album each month.

The problem is that royalty rates from streaming services have been consistently low. At one point Spotify was paying 0.5 to 0.7 of a cent per stream (or $5,000 to $7,000 per million plays) for its paid tier, and as much as 90 percent less for its free tier. By comparison, a typical artist used to earn 7 to 10 cents on a 99-cent download, after deducting amounts for the retailer, the record company, and the songwriter.

Between the artist and the consumer are businesses making significantly more revenue than the artist. With new technology disrupting old-style businesses in almost every other sector, what will it take for distribution technologies to close this gap? Rock bands knew about songwriting and having their own record labels way back in the dinosaur age. Bob Dylan decided he was a "folk" too and could write his own songs, and proceeded to graft new lyrics onto old folk tunes. The Beatles eventually became their own record company (Apple Records), the Rolling Stones started their own label, and many others sought to close the gap between the artist and the consumer.

The key to increasing these royalty streams has always been for the artist to write the songs and *be* the record company. The key to increasing exposure while also earning a living as a musician is to play live. Bands not only book their own live shows and handle their own publicity and social media connections, but also rent out venues themselves and charge admission. Some bands put on their own festivals. But it's a hard road, requiring business sense that artists are not known for having.

Nevertheless, the current wisdom is for artists to make use of these streaming services to get their music out there. The worry for successful acts is that streaming services cannibalize paid downloads and CD sales, and until there are enough streaming subscribers, they will lose potential revenue. History may be repeating itself, though. When the CD was first introduced, royalty rates were lower for CDs than for vinyl records. It took a while for CDs to become mainstream and replace vinyl for the royalty rates to rise.

The world of public-facing APIs has opened up the possibilities for modeling the musical influences of our favorite music artists. In 2010 we envisioned a project that would combine a crowdsourced database of musical entities linked to their influences, with a human interface like an inverted tree. Called the Rockument History Tree (partially developed in 2020), it would let you find out where an artist or band's musical ideas came from, and follow the trail of a musical concept backward or forward in time, essentially assembling a timeline of a band and its influences and followers. You could find other artists or bands that covered a particular song. You could even find other artists or bands from the same scene or birthplace, or who used the same producer.

The Beatles on the Rockument Tree:

Members/contributors	Partners/collaborators	Record Labels
John Lennon	Rory Storm and the Hurricanes	Apple Records
Paul McCartney	Billy J. Kramer	EMI
George Harrison	Tony Sheridan	Parlophone
Ringo Starr	Roy Orbison	Ocean Records
Stuart Sutcliffe (early member)	Gerry and the Pacemakers	Swan Records
Pete Best (early member)	Rolling Stones	Vee-Jay Records
Billy Preston (contributor)	Animals	Polydor Records
Klaus Voorman (contributor)	Beach Boys	Capitol Records
Andy White (contributor)	Bob Dylan	
Mick Jagger (contributor)	Ravi Shankar	
Keith Richards (contributor)	Byrds	
Brian Jones (contributor)	Chris Blake	
Nicky Hopkins (contributor)	Jackie Lomax	
Eric Clapton (contributor)	Yoko Ono	
London Symphony Orch. (contributor)		

Side Projects	Historical Influences	Influenced Others (by producer)
A Hard Day's Night	Buddy Holly	George Martin
Help!	Carl Perkins	Phil Spector
Magical Mystery Tour	Elvis Presley	Geoff Emerick
Yellow Submarine	Chuck Berry	Glynn Johns
Let It Be	Eddie Cochran	
The Dirty Mac	Little Richard	
Plastic Ono Band	Smokey Robinson	
In His Own Write/a>	Everly Brothers	
A Spaniard in the Works	Lonnie Donegan	

Influenced Others (by year):	Influenced Others (by sub-genre):	Influenced Others (by location)
Before 1962	British Invasion	Liverpool (birth place)
1962-1963	Rockabilly	Swinging London (scene)
1964-1965	Rock 'n' roll	
1966-1967	Skiffle	
1968-1969	Folk Rock	
1970-1971	Hard Rock	
1972-1973	Heavy Metal	
After 1973	Pop	
	Soft Rock	
	Singer-songwriter	
	Garage	

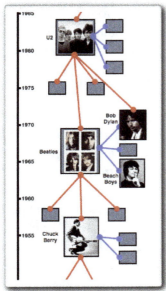

The Rockument History Tree project.

It is an ambitious project just to create the tree structure, starting with the Beatles, Bob Dylan, Rolling Stones, Ray Charles, John Coltrane, and James Brown, and work our way back to the beginning of recorded music, with the help of selected music historians. Moving forward, we want to invite living members of bands to add themselves to the tree, and eventually open it up to the world's bands in a controlled frenzy of crowdsourcing.

At the very least, it would be a crowdsourced method for artists and bands to place themselves in historical context for promotion purposes, and a handy tool to update their Wikipedia entries. But the goal is to track our musical heritage. Before music dies, we want to capture our collective intelligence about it, and put together the playlists that would define music history for future enthusiasts.

Unfortunately we have not mustered the resources to further develop this project. We pitched the idea to music industry experts who couldn't see how to make a profit from it. And so the rumbling distribution side of the music industry lurches on, overdue for a disruption and still blocking the lanes of progress.

An experiment

We enjoy books about music, artists, and bands. My particular eccentricity is to read lots of biographies and discographies, and to get a feel for how the music was created. Several outstanding books come to mind that truly define what it means to be a rock music lover in the latter half of the 20th Century: *Mystery Train: Images of America in Rock 'n' Roll Music* by Greil Marcus, *It Was Twenty Years Ago Today: An Anniversary Celebration of 1967* by Derek Taylor, *Last Train to Memphis: The Rise of Elvis Presley* by Peter Guralnick, and *The True Adventures of the Rolling Stones* by Stanley Booth.

When I read books that refer to music, I always want to play the music in order to understand the context. When I decided to try my hand at writing the "great American novel" (a theme among 20th-century American writers), I decided to include music as an essential part of the novel, adding another dimension of irony and commentary.

My novel, *The Experiment*, is about the power of music and the awesome responsibility of musicians to assimilate the vibrations of the universe. With streaming music, you can tap a link in a footnote, and hear the music while reading. The Apple Book version of my musical novel links directly to the Apple Music library. The book's website also offers Apple Music and free Spotify playlists. The book starts off with the very first song ever written, from ancient Sumer, performed by a historian-musician.

The novel starts with an introduction to the power of music in West African culture of the early 19th century, hinting at ancient secrets that defy the erosion of time and offer immortality to those who can wield it. It then tells the story of three young men growing up in the Sixties who are inspired by what they discover in this story — about the roots of rock 'n' roll, the power within the blues, the African spirits, High John the Conqueror, mojo, and the true meaning of the rolling stone. I was inspired by modern novelists. Thomas Pynchon, Don DeLillo, Ishmael Reed, Norman Mailer, Ken Kesey, Richard Farina, and Kurt Vonnegut are but a few.

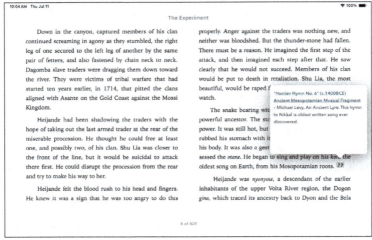

A footnote in the Apple Book version of The Experiment *by Tony Bove (on an iPad), using a footnote (top) to link directly to an Apple Music playlist (bottom) with "Hurrian Hymn No. 6", an ancient Mesopotamian musical fragment.*

Of all the projects described in this book, this is the one closest to my heart and soul, and the most successful in terms of reviews. Here's one from Rodney Osborne, author of the historical novel *Straw Men*: "Bove has written a huge, sprawling novel that all ties together around an obvious love for music and a fascination with a particularly important time in the history of music. I don't think I had ever thought of there being a 'history of music' before reading this novel. It takes you places and gets you thinking about how music develops and how it is passed down from one generation to the next in a way I had never considered before."

Another review, by Michael Gosney (Digital Be-In and Synergetic Press), caught the cleverness of the media combination: "The Apple Books edition embeds links to the music that are so important to the story that it's hard to imagine a better use of e-book technology."

Tools for mixing playlists

In this section we list the tools we used for creating, mixing, producing, and playing music and other projects, such as electronic books, that use music.

Media: Radio, vinyl records, tapes, CDs, MP3 digital music, iTunes downloaded or purchased albums and songs, Apple Music, Spotify

Tools and languages:

Apple Music (iTunes), iPod, MusicMatch, Windows Media, GarageBand, Pro Tools

Digital instruments, digital tuners, Cleartune, "Bullet" microphone

Apple Pages, Apple Books, Kindle Publisher, Adobe Photoshop

Surviving New Media

Our knowledge can only be finite, while our ignorance must necessarily be infinite. — Karl Popper

Hoare's Law of Large Programs: Inside every large program is a small program struggling to get out.
— Murphy's Computer Law

Rhodes' Corollary to Hoare's Law: Inside every complex and unworkable program is a useful routine struggling to be free.
— Murphy's Computer Law

In 2019, a gunman killed 51 people in an attack that he live-streamed on Facebook from a head-mounted camera. The immediate and viral spread of this video was one, albeit horrific, example of how people using social media changed our world. Five years later mass shootings seem to occur every week, self-serving conspiracy theories drive our polemics, and fake news and videos divert everyone's attention away from our real problems.

How could we be so easily blindsided? We had been advocating new media technologies for decades, arguing media power to the people. We had made our living helping others get through and profit from the transitions from print to digital to online media. And yet, right before our eyes, history continues to repeat itself, with more mass shootings, not because technology has pushed us to the brink of madness, but because we seem not to remember history.

As humans, we have regressed to an era of mass ignorance. But we weren't attacked by *Terminator* machines. We did this to ourselves and to future generations by not investing heavily in education.

We may think we are smart because we use smartphones and smart watches, but we trade privacy and security for the temptations of convenience. In order to have all the media we could possibly want, all of the time, and anywhere we want, we are willing to become targets for a bombardment of advertising and disinformation. Influencers tell us what to buy, how to act, who to like, and who to vote for, and we either do not know or do not care to admit that they are selling on behalf of others. We are lied to, *en masse,* for profit.

This is the tradeoff we make whenever we ignore fake news, or delegate decision-making to robots, or turn over our music and video choices to the profiteers in social media, or allow our vehicles to drive by themselves, or rely on spoken instructions from a GPS to get anywhere, or use ChatGPT to write our resumes. This is the tradeoff we make when we put occupational training ahead of liberal arts, and don't train our young to read and write.

As a result, we don't believe the actual news. Our children don't learn analytical thinking, and therefore don't build confidence in their own decisions. Our watching and listening experiences are vapid and meaningless. Some of us may never learn how to drive, or how to read a paper map. And we lose our personalities in the employment process as machines sort our work histories by keywords and, often, into the trash.

The spoken and written words, and even the videos, produced by our leaders and influencers can no longer be trusted. Frauds and scams dominate our emails and messages. Fake media has surpassed real media by leaps and bounds. Social media is blamed for eroding our intelligence. How do we survive new media?

Can we fix stupid?

Comedian Ron White does a comic bit about how you should never marry for looks alone. Beauty problems can be fixed, but... "You can't fix stupid. There's not a pill you can take; there's not a class you can go to. Stupid is forever."

Social media networks are not the cause of the erosion of intelligence. Delete Facebook and something else will take its place. (Ironically, those who deleted Facebook were using Facebook to announce their decision.) The critics of social networks remind us that people are victimized by social attacks, while defenders say that people are just too gullible and sensitive. Neither is completely true.

Many intelligent people, especially those of a certain age (like ourselves), are mindful of history and wary of volunteering personal information. We don't have time for lifestyle question-naires. Like many of you, we don't fill our social media profiles with nutty philosophies and pictures of ourselves holding auto-matic weapons. We hope to not be bait for the trolls who work for the mind-spammers.

We are wise enough to understand that targeted marketing is a fact of life. So are data breaches. Bombarded as we all are by misleading ads, scams, and fake news, somehow we can tell the difference. Many of us are avid readers and have learned to treat news videos and headlines as light entertainment. We don't settle back to watch conspiracy videos, and we don't read the "news" stories from sketchy sources our backwoods or urban cousins post in their fits of rage. We like to think that we harbor a healthy skepticism about everything, and always follow the money to get to the truth.

How did we get to such a state of grace, in the onslaught of the noxious stupidity that has overthrown intelligence in politics, entertainment, news, you name it? Our bullshit detectors were forged in our youth, in our well-rounded, liberal arts education. The more we absorb culture, and the more empathy we show toward humans of other cultures, the more we spread a sense of reasonableness about the entire world — and when something falls outside of the realm of reason, we reject it as false.

We hope Ron White is wrong and that we *can* fix stupid. We need to resist our gullible nature. We need to know more about tribalism so that we can understand why people accept hypocrisy among their own. And most importantly, we need to take a long view of education — like the way we think about global warming. The next generation may not have a sustainable planet, but why would it matter if their minds are melting from ignorance? A well-rounded education is just as important as a well-cared-for planet.

Amplified hate and deception

The pace of change is so rapid, and its impact so deep, that human life has already been transformed by deceptions of a grand scale. Lies and fallacies are growing exponentially, out of control, and possibly irreversible, aided and compounded by our merciless addiction to media, and by an access explosion that is erasing the boundaries between the first, second, and third worlds. As a result, the loudest liars and fallacy promoters are able to expand their control. This is why the grossest and least trustworthy politicians are able to remake themselves into autocrats.

The unexpected outcomes of the 2016, 2020, and 2024 American presidential elections sparked a renewed interest in the issue of fast-spreading false information, rumors, and memes, widely circulated on social media. Since the 2016 election, the term *fake news* in this country has come to represent an array of misleading news-style stories that were fabricated and promoted on social media to deceive the public for political, ideological, or financial gain.

Fake news includes reports that are intentionally false, and hoaxes created for the intent of going viral on social media. Malicious actors create multiple accounts on social media and fire away with multiple messages to boost proliferation and stick in people's minds. The echo-chamber effect takes over to reinforce these messages in social networks using celebrities and influencers who have millions of followers.

George Orwell predicted how technology could be used to fake reality, including fake news (as "Newspeak") in his last novel *1984*, written in 1949. Consider this quote from the novel, in the context of today's X (formerly Twitter): "Newspeak was designed not to extend but to diminish the range of thought, and this purpose was indirectly assisted by cutting the choice of words down to a minimum." Thus, tweets on X (formerly Twitter), with their minimum number of characters and abrupt abbreviations, have indirectly assisted in the spread of false information.

Orwell managed to capture the feeling of being held captive by a false reality: "Always in your stomach and in your skin there was a sort of protest, a feeling that you had been cheated of something that you had a right to." And you *have* been cheated.

Media began as a tool to enrich your life, and it should still be one today. Books, newspapers, magazines, radio, records, television, videos, and even websites were dedicated to providing content that could educate and entertain. But how much of it now is pure pap and not worth your time to watch, listen, or read? Even worse, as Orwell pointed out, how can you tell how much of it is a lie?

It may be true that you consider yourself better off now than before the digital media revolution. And yet, maybe you feel a mute protest in your bones. As Orwell put it, the instinctive feeling that the conditions you live in are intolerable, and that at some other time these conditions must have been different.

Deep fakes are not new

One of the oldest examples of fake news is a reference to the Greek philosopher Socrates, who is quoted on the prevalence of lies and exaggeration as a form of deceitfulness for public gain, or "falling under the spell of an orator". We surmise that Plato got the dimensions of Atlantis wrong when he transcribed Egyptian records, which is why he placed Atlantis in the Atlantic Ocean. During the first century BC, Octavian, emperor of Rome, ran a campaign of misinformation against his rival Mark Antony, portraying him as a drunkard, a womanizer, and a mere puppet of the Egyptian queen Cleopatra VII.

Fake news led the charge for all sorts of discrimination, including anti-semitism. In Trent, Italy, on Easter Sunday, 1475, a child had gone missing, and a Franciscan preacher gave a series of sermons claiming that the Jewish community had murdered the child, drained his blood, and drunk it to celebrate Passover. Fake news played a role in catalyzing the Enlightenment when the Catholic Church's false explanation of the 1755 Lisbon Earthquake prompted Voltaire to speak out against religious dominance.

In America, Benjamin Franklin wrote fake news about murderous "scalping" Indians working with King George III in an effort to sway public opinion in favor of the American Revolution. In the Great Moon Hoax of 1835, The New York Sun published articles about a real-life astronomer and a made-up colleague who, according to the hoax, had observed bizarre life on the moon. Taking fake news into the realm of fiction, in 1938 actor and filmmaker Orson Welles broadcast an adaptation of H. G. Wells's novel *The War of the Worlds* as a series of simulated news bulletins ("The War of the Worlds"), causing mass panic in New Jersey.

More recently, fake news, misinformation campaigns, and social media frenzy led to a sort of hysteria that resulted in people being killed. In 2016, an absurd election conspiracy theory circulated that Democratic presidential nominee Hillary Clinton allegedly led a child-trafficking ring out of the Comet Ping Pong bar and pizzeria in Washington D.C. In what is now known as the #Pizzagate shooting, a 28-year-old man walked into the place and fired shots from his AR-15 assault rifle.

While there have always been crazy people and mass murderers looking to be mentioned in history books, social media has amplified this tendency to become famous by shooting up a church or school Shootings every week make us inured to violence. A constant barrage of photos and videos of death and destruction have made us jaded to war. The old Lettrist International, with its doctrine of divertissement that led into the Situationist movement, would feel justified by the millions of viewers today obsessed with idiocy and fake news, with hundreds of thousands of perpetrators all over the globe jockeying for position as running propagandists in the race to beat democracy to the ground. Was this the "new idea" they had talked about? The dissolving of all seriousness in politics, the drowning of true art in a sea of nonsense?

The disseminators of fake news, deep-fake images, and misinformation are borrowing a tactic used by the Situationists in the 1960s, which focused on the concept of the *spectacle*, a unified critique of society that attempted to change reality by expressing and mediating social relations through images. Where there are fires, the false-idea disseminators carry gasoline to ignite spectacles. They mean to turn our words back on ourselves, forcing new speech even out of the mouths of the guardians of the good and right. It's an aesthetic occupation of enemy territory, a raid launched to seize the familiar and make it into something else. This is how liberals have been branded as radicals by the truly radical right who've turned conservatism on its head.

The key to creating fake news, misinformation, and disturbing memes is to misappropriate words and pictures, divert them into familiar scripts and theories, and blow them up. Every yes becomes a no, every truth dissolves in doubt, and everything changes. The freedom to say anything becomes the freedom to *do* anything. Crowds gather around every dialogue; dramas are enacted in every corner of the internet.

Art no longer has anything to say about it; has in fact joined in the fun. The role of art in society is to express our inner thoughts, feelings, and experiences, and act as a collective memory that shapes and steers society in ways we can't even describe. But art seems to have burned itself out in a war against its own limits, in a struggle to escape museums and amusement parks. What passes as works of art are imitations of ruins, set in a dismal yet profitable carnival in which each cliché has its disciples and each regression its fans.

The consummate modern piece of art is the half-shredded Banksy painting, an image on a white canvas in a frame containing a hidden shredder. As the picture is unveiled at the gallery, the canvas is supposed to drop down and be completely shredded, destroying the art. However, the gallery owner wisely stopped the process midway, leaving half-canvas and half-dangling shredded pieces, which is infinitely more profitable than a heap of shredded mess.

Banksy painting half-shredded (photograph by Sotheby's).

After many decades of various revolts against institutions and governments, we feel as if we've been sleepwalking through decades of high-tech fantasies. Now, in part due to the internet of many voices, trust — in institutions, governments, media, business, and organized religion — has eroded. The revolutionaries of the past wanted to open up the media and give voice to the millions of the proletariat. Now that they have their wish, with millions now publishing something every day, the media is overwhelmed with babble and run by companies that want to continue being profitable no matter what happens to the world.

To recognize truth, the only reliable instruments we have are our thoughts and our history. These instruments are coming together in a controversial way in the form of artificial intelligence (AI).

The machine is only learning

Will AI come to our rescue? No, technology never rescues anything, but it can be put to good use.

Can we use AI to help understand the complexity of the world around us? Perhaps, but only in the backhanded way of showing us how flawed our information universe is.

Can we at least use AI to detect fake news? MIT researchers believe that the best approach is to focus AI efforts on the news sources, not the individual pieces of content. They've demonstrated an AI system that determines if a source is accurate or politically biased. It needs only about 150 articles to detect the factuality of a news source, which means it could be used to help stamp out new fake news outlets before their stories spread too widely.

Sites like Snopes have traditionally focused on specific claims, which is admirable and believable but tedious; by the time the sites have verified or debunked a fact, there's a good chance it's already traveled across the globe and back again.

What we want is some intelligence between us and the internet that can automatically root out fake news *before* it appears.

That intelligence used to come from human editors who would curate the content, separating the real from the fake. You could trust the news from certain organizations that had proven themselves over time to report the truth, and you could compare several sources to figure out the bias in different accounts. But with AI, you have to put your faith in algorithms.

How bad could this get? Look at Microsoft's decision to replace its journalists with AI. In 2023, false stories and fake news paraded across the pages of the company's homepage, also known as MSN.com and Microsoft Start, which is one of the world's most trafficked websites and a place where millions of Americans get their news every day.

The simplest form of AI is the *model*, or collection of algorithms that can do something, such as enabling your phone to recognize your face. That model must be *trained* to recognize your face among many others, and that process — of feeding training data to the model — is called *machine learning*.

The capabilities of a particular model depend on much it has learned over time. As of this writing, AI is everywhere, underlying nearly all complex computer interactions and attempting to provide a security blanket for instant transactions and communications.

AI can also magnify human intelligence at top speed. Science-fiction movies have already taken us to the place where machines take over: Skynet in the *Terminator* series, the Architect in the *Matrix* movies, and the Cyclons of *Battlestar Galactica* are prime examples. Like living things, the machines create more versions of themselves so incredibly fast that humans don't notice.

Unlikely for sure, but AI is more pervasive than you may think. You are feeding AI models whenever you travel, purchase products, or make choices in a survey. You feed models directly whenever you use Google Assistant, Amazon Alexa, or Apple Siri, but you also feed them indirectly through everyday activities. AI systems measure your progress through traffic patterns, stores, and transit systems. AI tech recognizes not only faces on an iPhone but also fingerprints (and voiceprints), and is now widely used in medical applications.

How much can a machine learn, and how deep is that learning? If you take the sum total of human experience, expressed in words and images, and feed it into the machine, it shows us an artificial mirror of our world. The problem is, the mirror it shows us is cracked with false facts. We give this machine the responsibility for weighing the importance of one factoid over another, using algorithms to work with what it has already learned. But the machine can't fathom human emotions and feel the anger we feel about false facts. It puts together a narrative that can have rather large cracks in it.

We use AI to hold up an imperfect, cracked mirror of ourselves. AI is just a reflection of what we know, what we think we know, and what we don't know. It's the blind leading the blind. Garbage in, garbage out.

AI errors can be frustrating. As a rock music historian, I get irritated when I see on Facebook a picture of The Band mislabeled as the New Riders of the Purple Sage, or an article about Bob Dylan playing Woodstock in 1969 when he was nowhere near the show. I want future generations to get the correct information, even in advertisements.

We are all bombarded with AI-targeted advertisements. Ultimately our reaction is to distrust all advertising, in a similar manner that the proliferation of lies causes us to distrust all politics. Similarly, the spread of AI-generated deep fakes will eventually make us distrust AI. We may reach a point where believing that the results from AI are correct becomes an article of faith, a new religion: causing us to distrust reality itself.

Generating stuff with AI

The fantastic rave-up about AI is a bubble, just like the Web 1.0 craze of websites devoted to text, images, and videos was a bubble. But the bubble burst. Eventually the content websites became essential for business and pleasure and were absorbed into the fabric of the internet and the Web 2.0 services. Same with AI: The capabilities we are raving about will burrow into the fabric of everyday apps and services, and surface as new features. The one exception may be generative AI: instructing AI to generate the text, images, and videos of media.

Our more specific interest in tools such as OpenAI's ChatGPT is the ability to use it for formulaic writing. This is the literary side of what is called *generative* AI. In this manner you could, for example, compose poetry or wedding vows by using the tool to accumulate and condense web browsing into an appropriate choice of words. But the tools stray into incompetence territory when you try to use them to write law briefs or technical documentation.

Visions of the Terminator notwithstanding, we were not afraid to try AI for writing assistance and, in particular, for producing highly formatted documentation. Many authors have blazed this trail using semi-automated editing tools like Grammarly, online dictionaries, and translators. The new large-scale language models (LLMs) are useful for light editing. As some have pointed out, it is difficult to detect whether any given text snippet was produced by a language model.

Although we avoid using AI, we don't have that problem with writing helpers. We allow words to be spelled for us, and we often accept AI-suggested words and phrases. These helpers exemplify how AI can disappear into the fabric of our work. They would reinforce laziness in a writer starting out in this world, but for someone as seasoned as we are, they are harmless.

In the course of my work writing about machine learning, I decided to take ChatGPT for a spin at producing documentation about a public open-sourced API. It turned out to be useful for examining and explaining sample Python code, but not accurate enough to rely on, and not capable (yet) of adding the context and meaning for understanding the code. Perhaps in another year that machine will have learned more about the API and about Python.

A tool such as ChatGPT gives you relevant answers only if you write a good *prompt*, which is a command or an action sentence that offers specific useful information. For example, "Write an itinerary for my family's 5-day trip to Rome making a list of 10 additional activities". Like search terms, the narrower you make them, the more relevant are your results. Thousands of websites and blogs cover this topic, but for starter examples, see *How to write better ChatGPT prompts*.

The big yawn

Generative AI tools have already learned everything they need to know in order to produce novels, poems, and songs, but just about all are of questionable artistic merit.

For example, a set of lyrics written by "J. Lennon/AI" were essentially cobbled together by ChatGPT relying on John Lennon's real lyrics. I wasn't at all interested in reading them, as it seemed like a waste of time. Only the human mind can create interesting content. Machines can spit out clever lyrics, but so what? They can't get beyond clever into the realm of emotion. Humans innately know that this is true. A piece of content is original, perhaps interesting, and even emotional, but only if a human created it.

Why is this? Because the creation of content is a human impulse to advance evolution. The only important evolution, to humans, is that of the human race. We actually *don't* want our machines to evolve beyond us because we are already afraid of them. Case in point: AI scientists are warning about an AI model that overruled a human shutdown order (see "The Moment We Lost Control" in *Swaine's World* by Michael Swaine, prominent historian, AI pundit, and poet.)

In order to move people emotionally, the machine would have to learn more than just how to trigger emotions and win critical acclaim. Even if ChatGPT or its equivalent could write a novel that would pass as written by a human, it would not be "novel" in the sense that it is something new that has sprung from a human brain. To write a real novel, the machine would have to think for itself.

As of this writing, AI doesn't think for itself. The algorithms draw on existing human writing and other forms of expression for historical perspective. What the machine knows is what everyone already knows — it's just that the machine can locate and put together relative answers faster than humans can.

So why would we want to read fiction written to formula, or nonfiction that we could not really trust? An AI-created song might be pleasant to listen to, and might even have a groove, but it's still a regurgitation of yesterday's human creativity. The songs generated by AI will still sound like songs that have already appeared in the entertainment market. And that's when the copyright lawyers show up.

Plagiarism at scale

Rapid progress often comes with unanticipated consequences as well as unanswered questions. There is, for instance, a question on whether works of art generated by models are considered novel or mere derivatives of existing work. That raises questions of ownership. If ChatGPT produces a song with Dylan-like lyrics and a David Bowie style of melody, does anyone get royalties? Can we simply rephrase Norman Mailer for a new novel? Can film studios reuse an AI-created actor in subsequent movies without paying the actor? This is more than plagiarism at a higher scale, this is plagiarism gone wild!

Without a doubt, the training involved with AI exceeds all notions of copyright as we know it. We can read all the content we want, extract ideas from what we read, and then use those ideas and even partial phrases in our own works, without infringing copyright. But we also spend time rephrasing the phrases of others, re-composing the musical notes, redrawing from a different perspective, and so on. At this point, AI can't be used to rephrase, re-compose, or redraw well enough to be convincing. The content spit out by AI may infringe multiple copyrights.

The arguments in favor of this "plagiarism at scale" without changing the copyright laws include the fact that if copyright imposes a new licensing system now, after the cat is out of the bag, it would cause chaos as developers try to identify millions and millions of rights holders, for very little benefit because the royalty amount would by tiny for each one.

Another argument is that generative AI learns the same way humans do. Eager to learn as youngsters, AI machines harvest knowledge from everywhere and put it to use. Publishers of large quantities of information don't want their libraries used as fodder for machine learning unless license agreements can be drawn up.

Legal experts have argued that if training could be accomplished without creating copies, there would be no copyright issue. Copying is just an intermediate step to extract the elements and does not reuse the copyrighted expression to communicate to others. Rather than replacing the expression, the AI model can create a wide variety of different outputs wholly unrelated to the underlying, copyrighted expression. By that reasoning, generative AI is a fair-use issue.

Billions have been invested in generative AI under the assumption that, under current copyright law, any copying to extract information as training for AI models is permitted. Besides, the growth in creative expression through generative AI may be precisely what the Copyright Act was intended to promote.

Restorations and hybrids

Generative AI is not the only form of AI for making art. Once again, as with music CDs and iTunes downloads, the Beatles have played an important role in validating the use of AI to *restore* art. Rather than generating content, AI was used to isolate the John Lennon vocal from the rough demo of an unreleased song he recorded decades ago. The remaining Beatles added their contributions and produced "Now and Then". Lennon's voice was extracted using the AI-backed audio restoration technology commissioned by Peter Jackson for his 2021 documentary *The Beatles: Get Back*. However, every sound you hear in the new song is a human-created sound, done intentionally by humans.

We're also seeing the beginning of a hybrid, in which an artist uses AI to amplify human creativity. One recent example is known as the "Fake Drake": Anonymous artist Ghostwriter went viral when the AI-generated song "Heart on My Sleeve", which mimics Drake and The Weeknd, was submitted for a Grammy.

A far more intense and controversial work of art is "Refik Anadol: Unsupervised", a twenty-four-foot-square, which is a constantly morphing abstraction on the ground floor of the Museum of Modern Art (MoMA) in New York City. AI generates the imagery, in real time, using as its dataset tens of thousands of pictures of works in the museum's collection. The algorithms blend the pictures to create something wholly original but in some imaginary space between them. The installation will continually show original images virtually forever, without ever repeating any. But there is a huge "so what?" factor that impedes the ability to assess this art, which is more like a light show. There is no context for understanding what you see, or relating to it personally. It is as if you were looking at a Warhol painting of Marilyn Monroe, but with no idea who Marilyn Monroe was.

Budding writers, musicians, and artists will expand their use of generative AI because their industries are very competitive. But the best writers may be crowded out by the incompetent. Generative AI is so useful for setting up the *process* for content creation that many of you who can type, and therefore think you can write, can create first drafts. AI is useful for formatting text into a structure, such as an essay, book chapter, or marketing brochure. The problem is that by handling the process for you, AI reinforces laziness. Subsequent generations will not know how to format a report, tell a story, or write a poem.

Even worse, the automatic production of marketing and product "literature" will clutter up our thoughts with false promises and jingles. The junk that marketers put out there grabs our attention and keeps us from addressing the very real problems that may ultimately destroy civilization. The generators of this material think people are actually reading it, but they are wrong. We are only reacting to images and jingles. We no longer trust marketing.

There will be no stopping the progress of AI, except perhaps for fine art such as painting. Younger generations are obsessed with getting things done faster. We all want to bake the cake as quickly as we can, but one of the most dangerous pitfalls of creating fine art is doing it too fast, before the art has had time to grow in quality. We forget that the process is where and how artists learn what they are doing and what it means. When fast and convenient produces dreck, or worse, a false sense of truth, will we ever believe what we see or hear?

AI for health and lifestyle

Technology makes a difference when it helps to save human lives. Personal computing, desktop publishing, desktop video, and searching the internet were activities that made significant changes and affected human lives, often putting people out of work in the process.

What we need is AI that can make a difference — for example, help us manage our health. We can envision features in our iPhones that can show us, at any time, a log of everything we ate over the previous week, the medicines we took, and whether any of those meds are in conflict. With AI it could suggest dietary supplements and exercises to burn calories. It would "know" the true state of a person's health by monitoring heart, blood pressure, breathing, and so on. Our doctor could download the information and use it. The point is that AI can be used to gather information without forcing us to enter it manually.

We already use AI for patient diagnosis and specialized medicine. AI algorithms can create individualized patient care and treatment plans by analyzing patient data, including medical history, vital signs, and lifestyle choices. An interesting starting point is remote patient monitoring. AI algorithms can continuously analyze patient data to identify subtle changes in vital signs or symptoms that may indicate potential health risks. By alerting healthcare providers promptly, interventions can be implemented at the earliest stage. For a concise description of what you can do with remote patient monitoring, see "AI in Remote Patient Monitoring: The Top 4 Use Cases in 2024".

Also interesting are the "world-building" AI tools that are used, according to Steven Levy at WIRED, "to create simulated and imagined worlds that understand the dynamics of reality... While current generative AI is language-based, [scientist Fei-Fei Li] sees a frontier where systems construct complete worlds with the physics, logic, and rich detail of our physical reality." This exploratory use of AI promises to provide realistic simulations for experiments, or even totally imagined universes, and will open up science to new possibilities.

Humans are the weakest link

For decades, humans have placed fast access and convenience at the top of their priority list, as demonstrated by the rise of the internet and smartphones, matched by declines in book, magazine, newspaper, and music CD sales. What we have mostly done is trade our privacy for all this speed and convenience.

Privacy, as it is conventionally understood, is only about 150 years old. Most humans living throughout history had little concept of privacy in their tiny communities. Sex, breastfeeding, and bathing were shamelessly performed in front of friends and family. When Abraham Lincoln tapped all telegraph lines during the Civil War, few raised any questions.

The lesson from 3,000 years of history is that privacy has almost always been a back-burner priority. Humans invariably choose money, prestige, security, or convenience when it has conflicted with a desire for solitude.

So how do you keep your personal information and identity intact? How do you protect yourself from phishing scams, info thieves, evil videos, and fake news? How can you keep your data safe and your mind from going insane?

Forget about going non-Google, picketing Facebook, or boycotting X (formerly Twitter). The world's databases already have information about you. Internet services already know what you bought and may want to buy again. We could list over a hundred techniques they may use to get your personal information, and that list would change overnight with each new method they invent. Trying to change this situation is like playing whack-a-mole.

The best offense is a good defense. You need to understand what the threats are, in simple terms, and how to identify and protect against them. Trying to patch together solutions for each attack vector is like trying to catch trapeze artists in free fall with bedsheets. For example, phishing is a huge challenge with many attack vectors, and you need a lot more than bedsheets — you need to understand how these vectors work.

Goodbye Mr. Phish

One way to do that is to meet Mr. Phish. His mission is to steal money. He doesn't look like a criminal, and his actions are mostly unseen and undetected. He makes his living by pilfering online credentials — usernames and passwords. His modus operandi is phishing: tricking people into logging into a fake website in order to capture their credentials.

And Mr. Phish knows that the vast majority of security breaches involve weak or stolen credentials. He frequently finds lucrative targets among corporate executives working for companies with seemingly lax security. He uses the stolen credentials to impersonate them and siphon money out of personal and corporate accounts. His tricks work because he continually updates them, staying one step ahead of everyday security technologies and practices.

One day he learns on the Dark Web a trick to bypass Office 365 protections. His first step is to replicate the log-in page for Office 365, and register a domain for it that closely mimics the domain for Office 365 (such as "account.office356.emai"). The next step is to send you a legitimate-looking email with a security warning and a link to reset your password. He hopes the email will fool you into clicking the link and entering your password on the fake log-in page.

With little or no training you can spot many suspicious elements in Mr. Phish's email. For example, the sender's name might include brand-related text, which would be unlikely for a true company employee. The subject text, typical of phishing emails, uses a generic "Email Security Team" that probably doesn't exist for the real Office 365. The body text has a few grammatical errors, also typical of a phishing scam. And the link goes to a low-traffic website — very unlikely for a true account notification from Office 365.

But Mr. Phish is not done. He uses WhatsApp to send a phishing message to your mobile device. WhatsApp is useful because a message with a link shows a preview of the website logo and page title, which Mr. Phish can easily fake to appear legitimate.

Another way to get your password is to use stolen credentials bought on the Dark Web for a site such as Quora. Mr. Phish searches Quora to see if you are active on the social media question-and-answer site. Sure enough, you like to answer questions about the Beatles, and this becomes your downfall, because like many people, you use the same password for other systems besides Quora. There are just too many systems and services to remember a unique password for each one.

How do you defend yourself? The Federal Trade Commission offers good consumer advice, such as *How to Recognize and Avoid Phishing Scams*. The key takeaway is that if something seems too good to be true, it most likely isn't true at all. Remember the following important clues:

- Scammers pretend to be from an organization you know. The Social Security Administration and the IRS are frequently invoked, but it could be a software company, or any service you have an account with.

- Scammers pressure you to act immediately. That is, before you have time to think. They might threaten you with arrest, deportation, a virus on your computer, or exposure to your friends of a fictitious porn habit you supposedly have.

- Scammers want you to pay in a specific or suspicious way. Sometimes they insist on using cryptocurrency, or wiring the money through Western Union, or using a payment app or gift card.

The truth is, honest organizations won't call, email, or text to ask for your personal information, like your Social Security, bank account, or credit card numbers. In any case, it's best not to click on any links. Instead, use your browser to contact the organization using a website you know is trustworthy.

Tools for surviving media

In this section we list the tools we used for surviving media.

Media: News, fake news, deep fakes, artificial intelligence (AI)

Tools and languages:

Fact checking: Snopes, PolitiFact, FactCheck.org

Deep fake detection: Detect Fakes, FakeCatcher

AI: ChatGPT (generative AI), Grammarly (editing AI)

Think for yourself

Technical writers gorge on questionable data, from product specs and marketing gibberish to support questions, in order to understand how a product works and how to properly describe it. To survive media, you must learn to analyze what you get from questionable inputs and sources, and put it together in your own mind. You have to follow the money, as Woodward and Bernstein did to uncover the Watergate corruption scandal.

Despite the gains of the technology sector, wage growth across all industries has been essentially stagnant for years. As tech culture infiltrates every corner of the business world, its hymns to the virtues of relentless work remind us of nothing so much as Soviet-era propaganda, which promoted impossible-seeming feats of worker productivity to motivate the labor force.

In the new work culture, enduring or even merely liking one's job is not enough. Workers should love what they do, and then promote that love on social media, fusing their identities to that of their employers. They should glorify personal profit, even if their bosses and investors — not workers — are the ones capturing most of the gains. It's not difficult to view this hustle as a swindle. Convincing a generation of workers to beaver away at their jobs is profitable only for those at the top. Generations raised to expect that good grades and extracurricular over-achievement would reward them with fulfilling jobs that feed their passions, instead wound up with precarious, meaningless work and a mountain of student loan debt.

According to Douglas Rushkoff, author of *Coercion*, the great "Napsterization" of economics in this digital environment has "made a bunch of billionaires and a whole lot of really poor, unhappy people." (See "Doug Rushkoff Is Ready to Renounce the Digital Revolution" in *WIRED*) And those billionaires are thinking for themselves. In his book *Survival of the Richest*, Rushkoff writes, "Instead of just lording over us forever, the billionaires at the top of these virtual pyramids actively seek the endgame."

The endgame, it turns out, is closer than you think. Climate change threatens the world's food supply. Vast stretches of land are turning into deserts. Racism and hypocrisy are on the rise. The super-wealthy know these things, but they persist in believing they are immortal, as if wealth brings immortality. And it seems to — they can afford every medical procedure, every type of assisted living. They are planning escape routes and building impenetrable monuments to themselves, such as large plantations in Hawaii and radioactive-proof bunkers in New Zealand. At the first sign of an apocalypse — nuclear war, a killer germ, a French Revolution-style uprising targeting the one percent — these billionaires plan to hop on a private jet and hunker down.

It is high time that you have choices of your own, so that when the end-times occur (or if the power goes out, or if you just can't stand all this tech nonsense and false information anymore), you can fall back on a basic survival lifestyle that doesn't require constant electricity and the internet.

Sharpen your basic mental skills to keep tech advances from eroding them. Read paperback books that are easy to carry. Put your keys in the same pocket all the time, so that you don't need technology to find them. Learn how to play music in your mind, so that you don't need any devices or headphones. Try playing each instrument in your mind, so you can truly experience the wonders of music and still react to the real world when you need to, without looking for a pause button.

Learn how to drive and especially how to park without relying on the car to do it for you. Use paper maps, those relics of a forgotten world, or even use the one on your phone, so that you don't have to rely on GPS and faulty step-by-step driving instructions that lead you up blind alleys, down dirt roads, and right into the lake. But whatever you do, don't act like your grandparents — don't ask for directions. Look at a map first, then feel your way and use common sense. Don't follow the crowd. Use intuition to find a good route — not necessarily the fastest but the best, the easiest, or the least stressful.

The following are a few more useful tips for dealing with social media:

- *Stop bleeding your personal information out to the world.*
  For example, resist the impulse to comment on what you like or don't like. Post only a few photos, if any, because you should assume that the photos will be copied and used for something else. Don't ever post photos of your children.

- *Trust news only from sources you trust.*
  Spot the media tricks to get your attention, such as click-bait headlines, and don't go back to those sites. Avoid ads that promise far more than they could possibly deliver.

- *Don't share information or videos you don't know to be true.*
  If you think something is false, it probably is. Check sites like Politifact and Snopes to investigate before sharing.

- *Look for the telling clues of nefarious actors.*
  Check for spelling errors, tortured grammar, and hyperboles in emails and messages, all of which are indicators of bad actors.

Use your common sense. Don't trust the crowd. Crowdsourcing can be a powerful way to gather information and disseminate casual learning, but it can also help to proliferate fake media before an election. Can you trust a crowd to produce something that is useful and accurate? Think of the situation as crowd-*trusting*. The question to ask yourself is, are there enough individuals in this particular crowd who know the real answers and can shout over all the others?

Follow this advice to spot fake media:

- Consider the source. Investigate the site, its mission, and its contact information.

- Does the picture make sense? In fake videos and pictures, think of the context and timeline. Would John Lennon really have hung out with Albert Einstein? Use your bullshit detector.

- Read beyond the headline. Headings and pull quotes can be outrageous to grab readers. What's the entire story?

- Check the author. Do a quick search to see if the author is credible, has admirers, has critics, or even exists at all.

- Check supporting sources. Click on those links to determine if the information given actually supports the story.

- Check the date. Reposting old news stories is a common trick to divert attention from real news that may be relevant.

- Is it a joke? If it is too outlandish, it might be satire. Consult a fact-checking website.

We should all take a cue from the high-schoolers who are marching for gun control: keep at it. Don't falter. These students are demonstrating that they know the difference between hard truth and NRA bullshit. We should all demonstrate for better education. We need to arm each new generation with history and common sense.

Above all, read the manual when you need to know how to use something. Don't just watch a video for instructions. It doesn't matter whether the documentation is online or on paper. You will find something useful in the words that you couldn't possibly catch in a fast-paced video. Someone took the trouble to write it down, and you can benefit from that work by reading it.

THE END.

About the Authors

Tony Bove and Cheryl Rhodes wrote more than two dozen published books on computing, desktop publishing, and multimedia. These include bestselling titles such as *The Art of Desktop Publishing, Free Software, Desktop Publishing with PageMaker, Adobe Illustrator: The Official Handbook for Designers, The Well-Connected Macintosh, Official Macromedia Director Studio,* and *iPod and iTunes For Dummies.*

They founded *Desktop Publishing/Publish* and *User's Guide* magazines and the *Bove & Rhodes Inside Report on New Media* newsletter, and wrote the weekly Macintosh column for *Computer Currents* for over a decade, as well as articles for *NeXTWORLD*, the *Chicago Tribune, NewMedia, Macintosh Today,* Prodigy Online, and the *Whole Earth Software Catalog.*

Tracing the personal computer revolution back to the 1960s counterculture, Bove & Rhodes produced a CD-ROM interactive multimedia documentary in 1996, *Haight-Ashbury in the Sixties* (featuring music from the Grateful Dead, Janis Joplin, and Jefferson Airplane). They also developed the Rockument music site, www.rockument.com, with commentary and podcasts focused on rock music history, and the *Tony's Tips* app for iPhone and iPad users.

Tony Bove currently produces developer documentation for Apple and plays harmonica in various Bay Area bands. Cheryl Rhodes provides technical assistance and media for a northern

California entertainment company, and works with elementary school students using augmentative communication and assistive technology tools to overcome learning roadblocks. For more information about this book and its authors, visit www.tonybove.com.

www.ingramcontent.com/pod-product-compliance
Lightning Source LLC
Chambersburg PA
CBHW052139070326
40690CB00047B/1138